Artur Juszczak

Mitsubishi A6M
ZERO

First published in Poland in 2001 by Robert Pęczkowski
Usługi Informatyczne, Orzeszkowej 2/57, 39-400 Tarnobrzeg, Poland
e-mail:rob   ertp@zt.tarnobrzeg.tpsa.pl
in cooperation with h
Mushroom Model Publications,
36 Ver Road, Redbourn,
AL3 7PE, UK.
e-mail:rog   erw@waitrose.com

© 2001 Mushroom Model Publications.

All rights reserved. Apart from any fair dealing for the purpose of private study, research, criticism or review, as permitted under the Copyright, Design and Patents Act, 1988, no part of this publication may be reproduced, stored in a retrieval system, or transmitted in any form or by any means, electronic, electrical, chemical, mechanical, optical, photocopying, recording or otherwise, without prior written permission. All enquiries should be addressed to the publisher.

WYDAWNICTWO DIECEZJALNE SANDOMIERZ

ISBN 83-7300-085-2

Editor in chief	Roger Wallsgrove
Editors	Bartłomiej Belcarz
	Robert Pęczkowski
	Artur Juszczak
Edited by	Robert Pęczkowski
Page design by	Artur Juszczak
	Robert Pęczkowski
Cover Layout	Artur Juszczak
DTP	Robert Pęczkowski
	Artur Juszczak
Translation	Roman Postek
Proofreading	Roger Wallsgrove
Colour Profiles	Artur Juszczak
Drawings	Dariusz Karnas
Printed by	Drukarnia Diecezjalna Sandomierz
Photos:	Arthur Lochte
	Robert Pęczkowski
	Roger Wallsgrove

Special thanks for

Ryan Toews

for providing drawings of Zero's stencils and help in camouflage research.

HISTORY

A6M2 of Koga under repair- July 1942, Naval Air Station San Diego. Please note removed engine and armament.

Photo A. Lochte coll.

Up to this date, the Zero fighter is considered the most popular and best known Japanese aircraft amongst air enthusiasts all over the world. Being the basic fighter used by the Japanese Imperial Navy in the Pacific War theatre, it was employed in all major aerial operations.

The Zero (for official Japanese Imperial Navy designation see table on page 6) was the first shipboard fighter capable of defeating its land-based opponents. Its introduction came as a shock to the Allies and numerous aerial victories in China and at the opening stages in the Pacific War, greatly contributed to the creation of the myth of a "super fighter" having no match among Allied ones. Indeed, the Zero was a formidable fighter aircraft at the beginning of the war. In time, however, its superiority began to diminish, so that by the end of the war it became an easy prey for the American fighters.

According to the doctrine officially employed by the Japanese military authorities, the main emphasis during the design of their new fighter has been on its maneuverability and maximum range, with appropriate armoured protection for the pilot and other vulnerable parts of the aircraft (i.e. fuel tanks) being of secondary importance. This vulnerability to the enemy's fire and inferior dive speed seem to be the major drawbacks of the Zero fighter. Excessively delicate

HISTORY

- and thus weak - wing covering was a limiting factor in attaining appropriate dive speeds. This situation was only corrected with the introduction of the A6M5 version.

The initial success of the Zero can largely be attributed to the fighter's great maneuverability as well as the immaculate training of its pilots. The best American fighter at that time, the F4F Wildcat, did not stand a chance against the Zero. American pilots flying Wildcats tried to avoid dogfighting the Zero at all costs. However, from the time of the Guadalcanal air battle the Wildcats began to engage in close-in dogfights with Zeros, and quite successfully so. The reasons for this shift in balance are many, but two seem to be the most obvious: a considerable decline in quality of the Japanese pilot training and the new tactics evolved by American fighters engaging the "super fighter". The main principle behind these was to always attack with the altitude advantage, and when in trouble, to escape by a sharp dive with great speed. From that time onward, the Zero fighter lost its previous reign and supremacy, and despite constantly implemented improvements and introduction of new versions, was never to regain it again.

Construction

Single-engine, low-wing cantilever monoplane of all-metal construction. Fuselage of semi-monocoque structure in two sections. Metal wings with two spars. Tail unit of classic cantilever design with fabric-covered rudder and elevator. Hydraulically retractable undercarriage.

A6M2s during take off from the carrier Shokaku to attack Pearl Harbour, 8 December 1941.
                                                                            Photo A. Lochte Coll.

Powerplants

Mitsubishi Zuisei 13	- fourteen-cylinder twin-row air-cooled radial, rated at 780 hp for takeoff.
Nakajima NK1C Sakae 12	- fourteen-cylinder twin-rHow air-cooled radial, rated at 940 hp for takeoff.
Nakajima NK1F Sakae 21	- fourteen-cylinder twin-row air-cooled radial, rated at 1130 hp for takeoff.
Nakajima Sakae 31	- fourteen-cylinder twin-row air-cooled radial, rated at 1130 hp for takeoff, with water/methanol injection.
Mitsubishi MK8P Kinsei 62	- fourteen-cylinder twin-row air-cooled radial, rated at 1560 hp for takeoff.

Armament:

2 x 20mm cannons Type 99 and 2 x 7,7 mm machine guns Type 97. (A6M1, A6M2, A6M3, A6M5, A6M5a)

2 x 20mm cannons Type 99, 1 x 13mm machine gun Type 3 and 1 x 7,7 mm machine gun Type 97. (A6M5b)

2 x 20mm cannons Type 99 and 3 x 13,2 mm machine guns Type 3. (A6M5c, A6M6c, A6M7)

2 x 20 mm cannons Type 99 and 2 x 13,2 mm machine guns Type 3. (A6M8)

2 x 30 mm cannons and 2 x 7,7 mm machine guns Type 97. (A6M3 - experimental)

3 x 20 mm cannons Type 99 and 2 x 7,7 mm machine guns Type 97 (A6M5)

Detachable armament

2 x 60kg in bomb load (standard)

1 x 250 kg bomb

1 x 500 kg bomb (maximum for A6M7 and A6M8)

8 x 10kg or 2 x 60kg unguided rocket projectiles (A6M7 and A6M8)

1 x 330liters - additional drop tank (all versions except A6M7 and A6M8)

2 x 350 liters - additional drop tanks (A6M7 and A6M8)

A6M2 Model 21 - foreground and A6M5 Model 52 - background taken at RAF station Seletar, Singapore, 1945. "ATAIU" - Allied Technical Air Intelligence Unit, "SEA"- South East Asia.

Arthur Lochte coll.

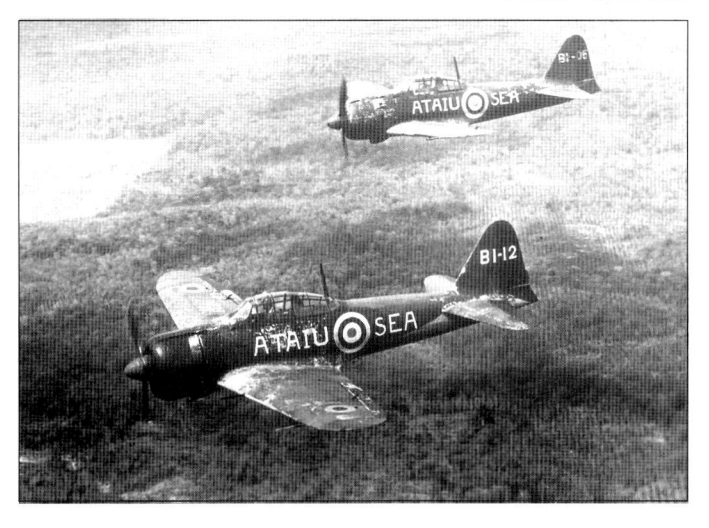

HISTORY

Version	Official Japanese Navy Name	Translation of Japanese Navy Name
A6M2 Model 11 – first 25 airframes	12 Shisaku Seizo Kanjo Sentoki	12 Experimental Carrier - Based Fighter
A6M2 Model 11	Rei Shiki Ichi Go Kanjo Sentoki	Type Zero Mark 1 Carrier – Based Fighter
A6M2 Model 21 (since airframe no 67 of all Zeros, wing with folded wing tip)	Rei Shiki Ichi Go Kanjo Sentoki Ni Gata	Type Zero Mark 1 Carrier – Based Fighter Model 2
A6M2 Model 21	Rei Shiki Kanjo Sentoki Ni Ichi Gata	Type Zero Mark 1 Carrier – Based Fighter Model 21 **(Revised name for the previous version)**
A6M3 Model 32	Rei Shiki Ni Go Kanjo Sentoki	Type Zero Mark 2 Carrier – Based Fighter
A6M3 Model 22	Rei Shiki Kanjo Sentoki Ni Ni Gata	Type Zero Carrier – Based Fighter Model 22
A6M5 Model 52	Rei Shiki Kanjo Sentoki Go Ni Gata	Type Zero Carrier – Based Fighter Model 52
A6M5 Model 52a	Rei Shiki Kanjo Sentoki Go Ni Ko Gata	Type Zero Carrier – Based Fighter Model 52a
A6M5 Model 52b	Rei Shiki Kanjo Sentoki Go Ni OtsuGata	Type Zero Carrier – Based Fighter Model 52b
A6M5 Model 52c	Rei Shiki Kanjo Sentoki Go Ni Hei Gata	Type Zero Carrier – Based Fighter Model 52c

A6M2, "DI-108" from the carrier "Ryujo" flown by Flight Petty Officer Tadayoshi Koga, crashed in the Aleutian Islands off the coast of Alaska on 4 June 1942. It was repaired to flying condition by the U.S. Navy at the Naval Air Station, San Diego, California in July 1942. Photo is of the test flight on 15 October 1942, piloted by either Lt. M.C. Hoffman or (more probably) Commander C.T. Booth.   *Arthur Lochte coll.*

VERSIONS
A6M2 Model 11

In July of 1940 the IJN accepted the model A6M2, and in August, serial production began at Aircraft Plant No. 3 of the Mitsubishi's factory at Nagoya. The engine destined for the serial fighter, the Sakae 12, was somewhat larger than the prototype's Zuisei 13, thus necessitating the enlargement of the engine cowling. Two profiled channels had to be implemented for the guns in the upper fuselage decking. Also, the carburetor air intake had to be moved to the bottom of the fuselage. With unit Serial No.22, the wings received a reinforced main spar. Production of this model ceased in November, after building 64 of the A6M2 Model 11 fighters.

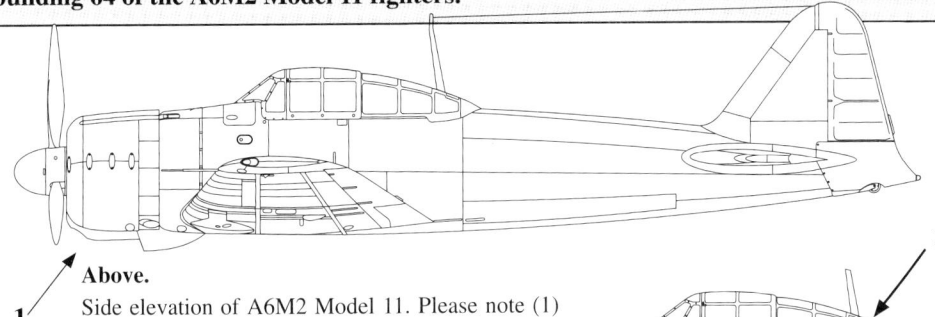

Above.
Side elevation of A6M2 Model 11. Please note (1) older version of the air intake.

Right.
Close up view of canopy used on Model 11 till airframe #47 with entirely glazed end piece. (2)

Stencil used on the first 25 Zeros with official designation 12 Shi Kanjo Sentoki

A6M2 Model 21 (s/n 3372) in Eglin Field, Florida, USA. Aircraft captured in China, rebuilt in USA by Curtis Aircraft and then scrapped later in the war.

Photo A. Lochte Coll

Mitsubishi A6M ZERO 7

VERSIONS

A6M2 Model 21

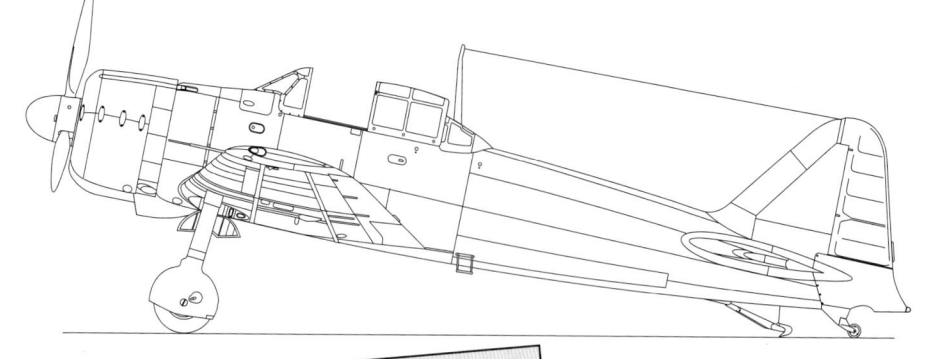

Upon successful completion of tests aboard the aircraft carrier Kaga, the new version was introduced into serial production - the A6M2 Model 21. A new feature was manually upward-folding wingtips, allowing the use of standard size aircraft carrier deck elevators. Beginning with aircraft No. 127, each aileron received a small tab balance in order to eliminate dangerous flutter. A total of 3368 aircraft of the A6M2 Model 21 was built.

Above.
Port elevation of A6M2 Model 21. Aircraft shown with undercarriage and arrestor hook in down position. Model 21 received mass balance under each aileron, but only 326 airframes had external mass ballance.

Below.
Wingtip in folded position. Wingtips were folded manually.

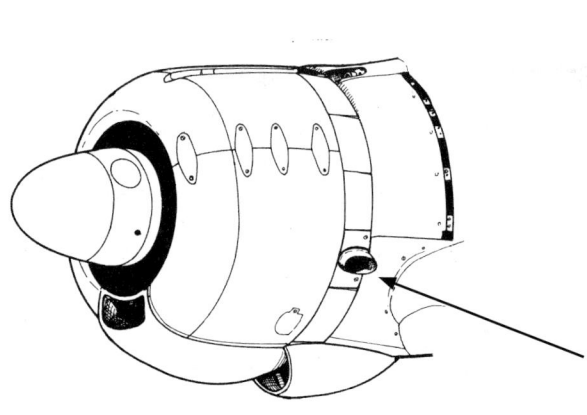

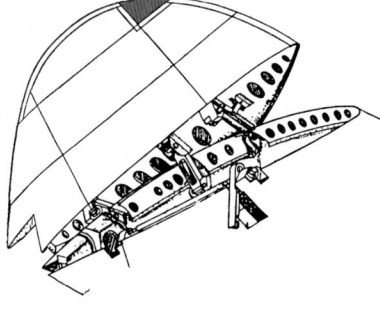

Left.
Up to airframe #34 of Model 11
1 exhaust tube exited from the fourth cowl flap.

8   Mitsubishi A6M ZERO

VERSIONS

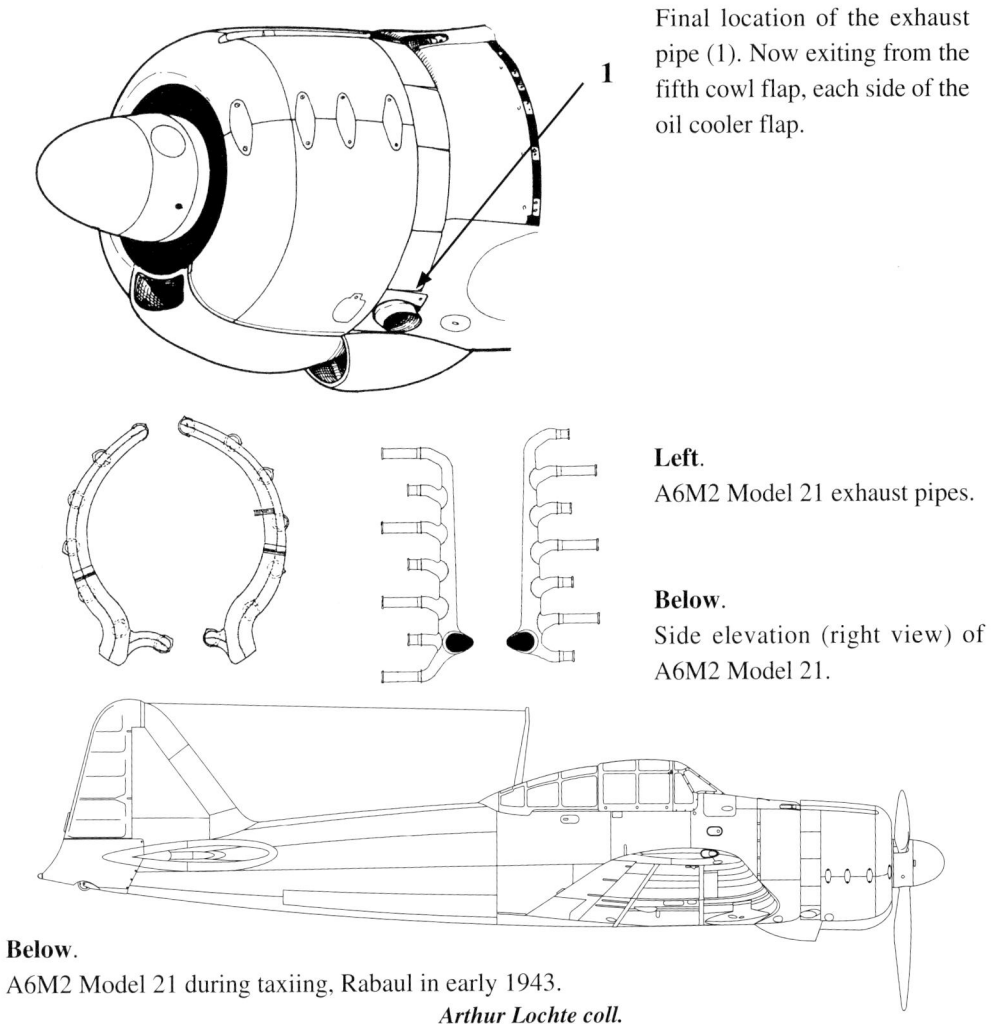

Final location of the exhaust pipe (1). Now exiting from the fifth cowl flap, each side of the oil cooler flap.

Left.
A6M2 Model 21 exhaust pipes.

Below.
Side elevation (right view) of A6M2 Model 21.

Below.
A6M2 Model 21 during taxiing, Rabaul in early 1943.
Arthur Lochte coll.

Mitsubishi A6M ZERO

VERSIONS

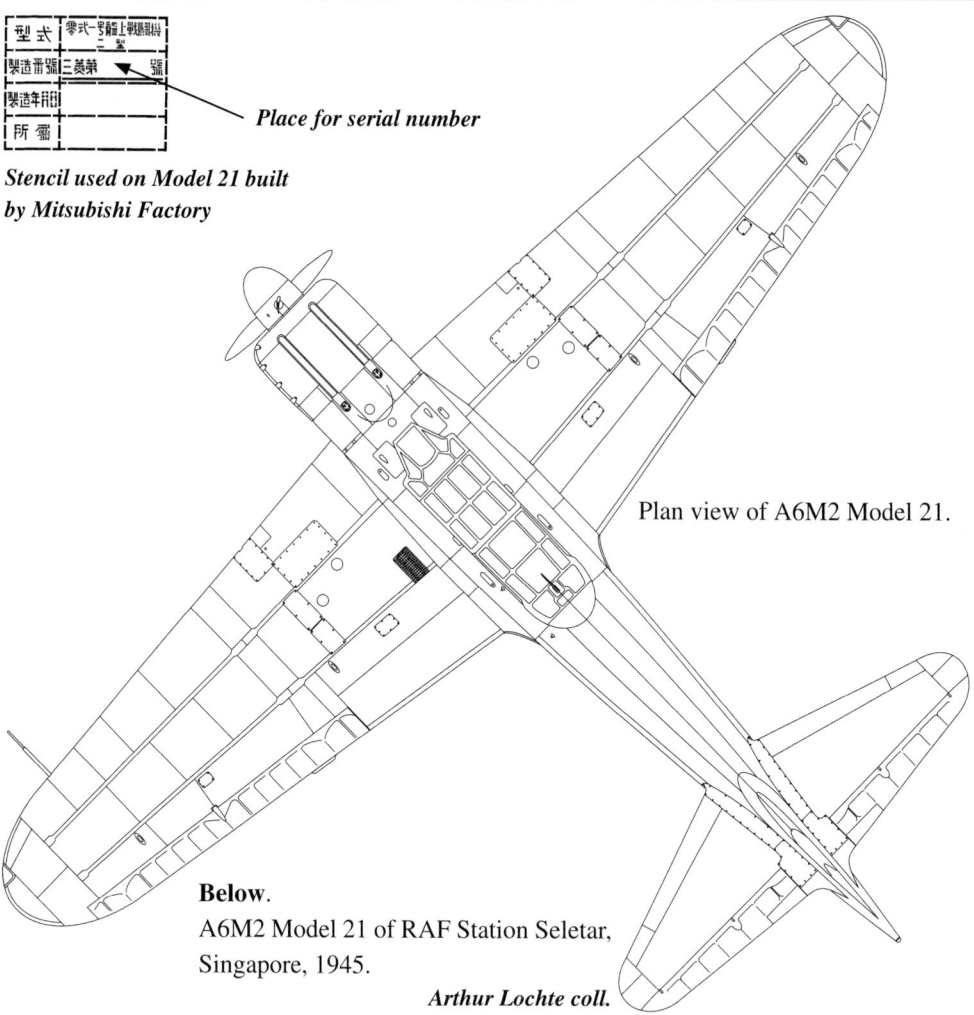

Place for serial number

Stencil used on Model 21 built by Mitsubishi Factory

Plan view of A6M2 Model 21.

Below.
A6M2 Model 21 of RAF Station Seletar, Singapore, 1945.

Arthur Lochte coll.

10 Mitsubishi A6M ZERO

VERSIONS

Stencil used on Model 21 built by Nakajima Factory till November 1942.

Plan view (undersurface) of A6M2 Model 21.

Above, right.
The Flettner tab (1) was used on the earliest Zeros but was thought to contribute to wing flutter. Consequently external mass balance tabs (2) were fitted (or retrofitted) to the first 326 Zeros. These in turn were replaced by internal balance weights. In the late model A6M2s manufactured by Nakajima the Flettner tabs were reintroduced.

VERSIONS

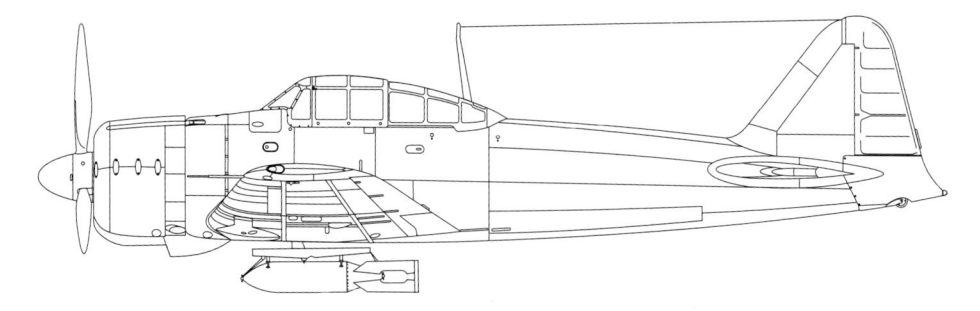

Above. Side elevation of Model 21 with 250 kg bomb.

Two photos of A6M2 Model 21 cowling. In the upper photo is bottom half with carburetter air intake. In the lower one is upper half of the cowling.

Arthur Lochte.

Only a few Model 21 Zeros were outfitted with long barreled 20 mm cannons and this was strictly a field modification that was not reflected by any change in the nomenclature. Late model Nakajima built A6M2s were, however, equipped with 100 round drum magazines instead of the earlier 60 round drums. Again, this did not result in any change in the aircraft designation.

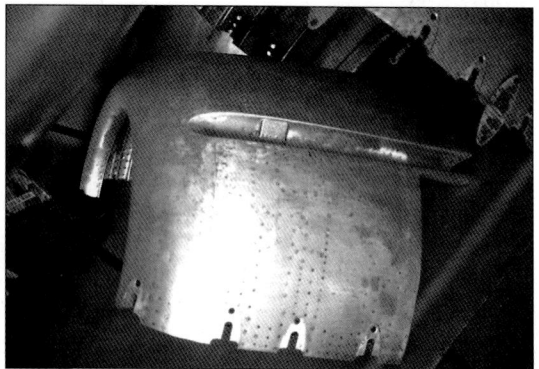

Below. Side elevation of Model 21 with modified wing armament.

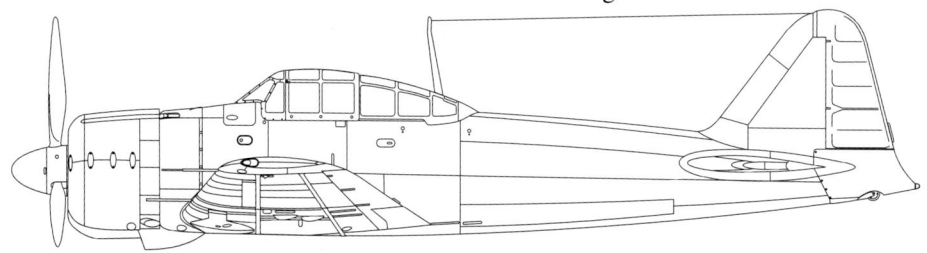

12 Mitsubishi A6M ZERO

VERSIONS

Left. Engine and spinner of A6M2 Model 21. Also bombs racks are visible.

A. Lochte coll.

Right. Front elevation of A6M2 Model 21. Aircraft with undercarriage in down position.

Stencil used on Model 21 built by Nakajima Factory since October 1942.

A6M2 Model 21 fighters, Kahili Airfield at Buin, Bougainville Island in April 1943. Fighters of Hiyo and Junyo carriers.

A. Lochte coll.

Mitsubishi A6M ZERO

VERSIONS

A6M3 Model 32

A6M3 Model 32 with clipped wing tips.
A. Lochte coll.

Work on this new model began six months before the outbreak of the war. The main goal was to achieve an increase in both the speed in level flight and the rate of climb. For this purpose a new engine was introduced - the Sakae 21, employing a two-stage supercharger as compared to the previously used single-stage unit. Due to the new engine's increased size and weight, some redesign of the fuselage structure had to be made. The firewall was moved 20 cm further aft, reducing the capacity of the fuselage fuel tank (from the original 98 liters to 60 liters). Also, the shape of the fuselage's front required reworking. The shaping of the supercharger air intake had to be modified. This new model also dispensed with folded wingtips. Its production started in July 1941, and ended after 343 machines were built.

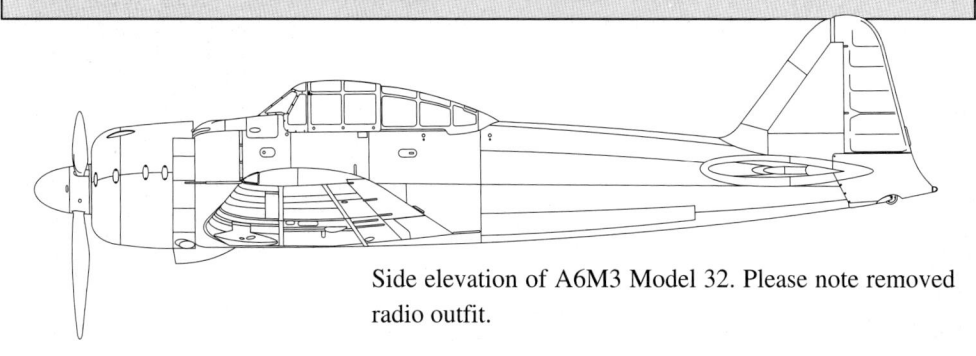

Side elevation of A6M3 Model 32. Please note removed radio outfit.

14   Mitsubishi A6M ZERO

VERSIONS

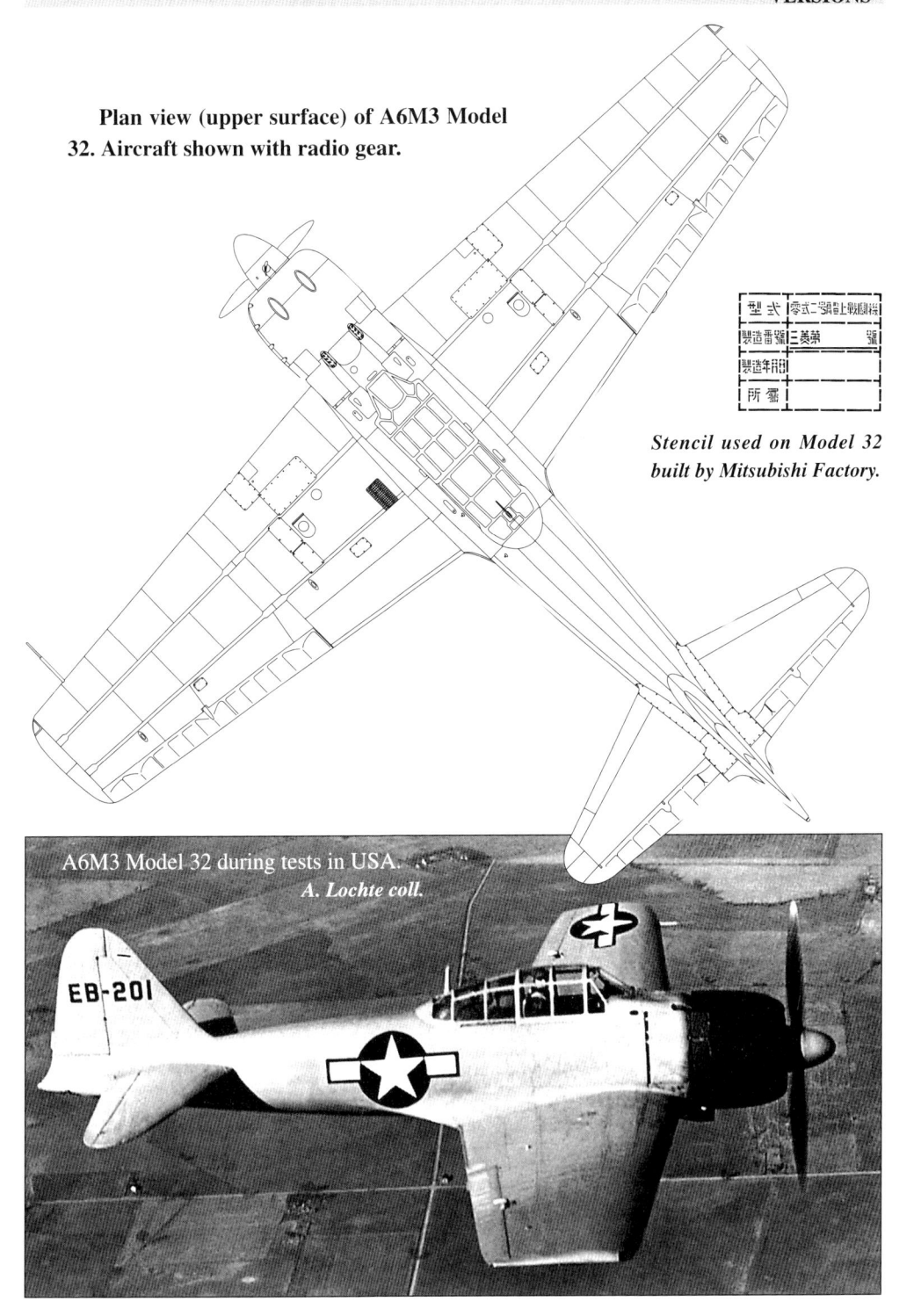

Plan view (upper surface) of A6M3 Model 32. Aircraft shown with radio gear.

Stencil used on Model 32 built by Mitsubishi Factory.

A6M3 Model 32 during tests in USA.
A. Lochte coll.

Mitsubishi A6M ZERO 15

VERSIONS

Right.
Drawing showing cannon Type 99 Mark 1 Model 3 - Kai with short barrel and drum magazine.

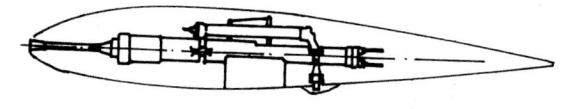

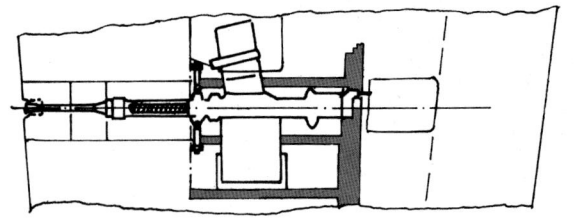

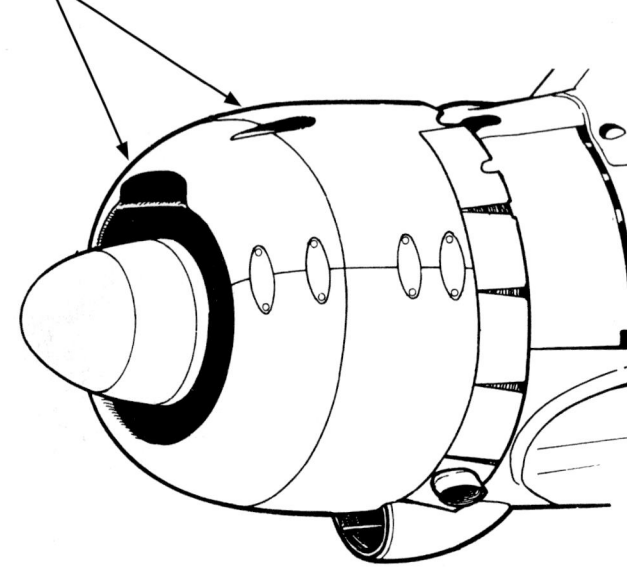

Left.
Drawing showing redesigned cowling of A6M3 Model 32. Please note relocated air intake and new gun outlets (1).

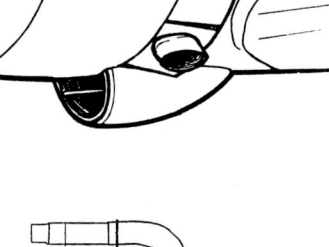

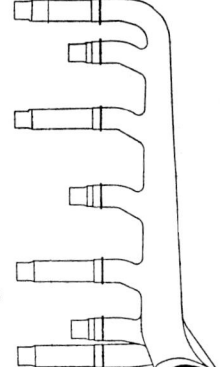

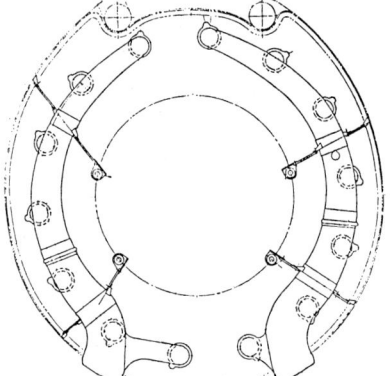

Right.
Drawings showing A6M3's exhaust pipes arrangement.

16 Mitsubishi A6M ZERO

VERSIONS

Below.
Front view of A6M3 Model 32.

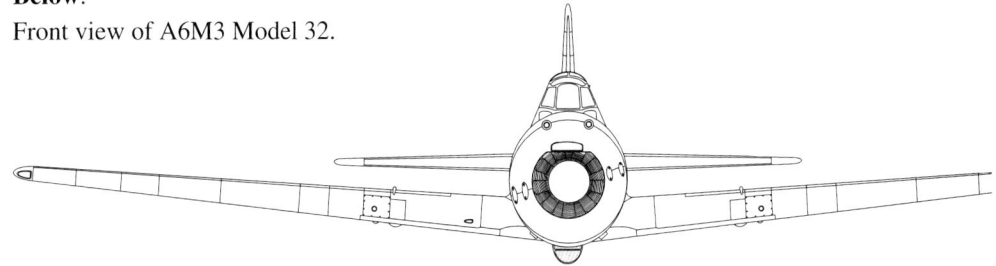

Right and below.
Two photos of A6M3 Model 32 captured by Americans end test flown in USA.
Both photos A. Lochte coll.

Mitsubishi A6M ZERO **17**

VERSIONS

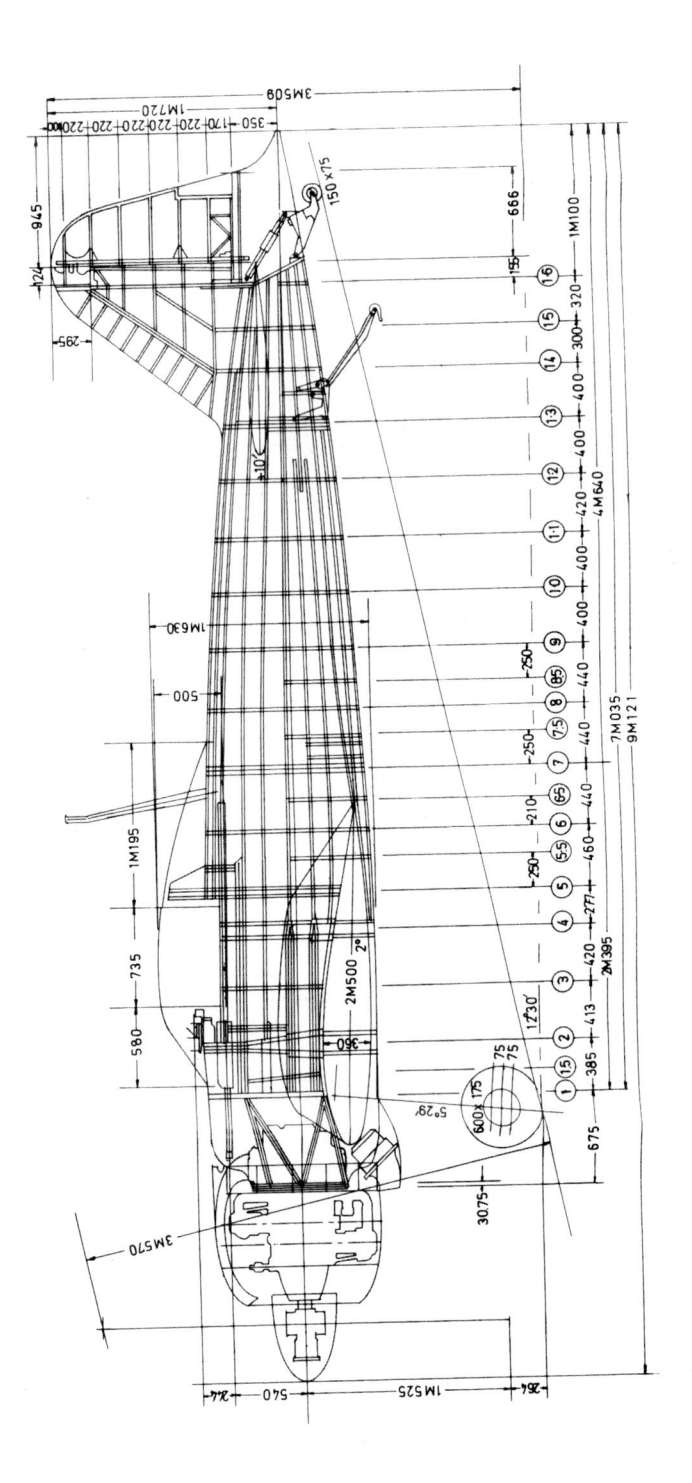

Fuselage dimensions of A6M3 Model 32.
Drawing from original technical Manual.

VERSIONS

A6M3 Model 22

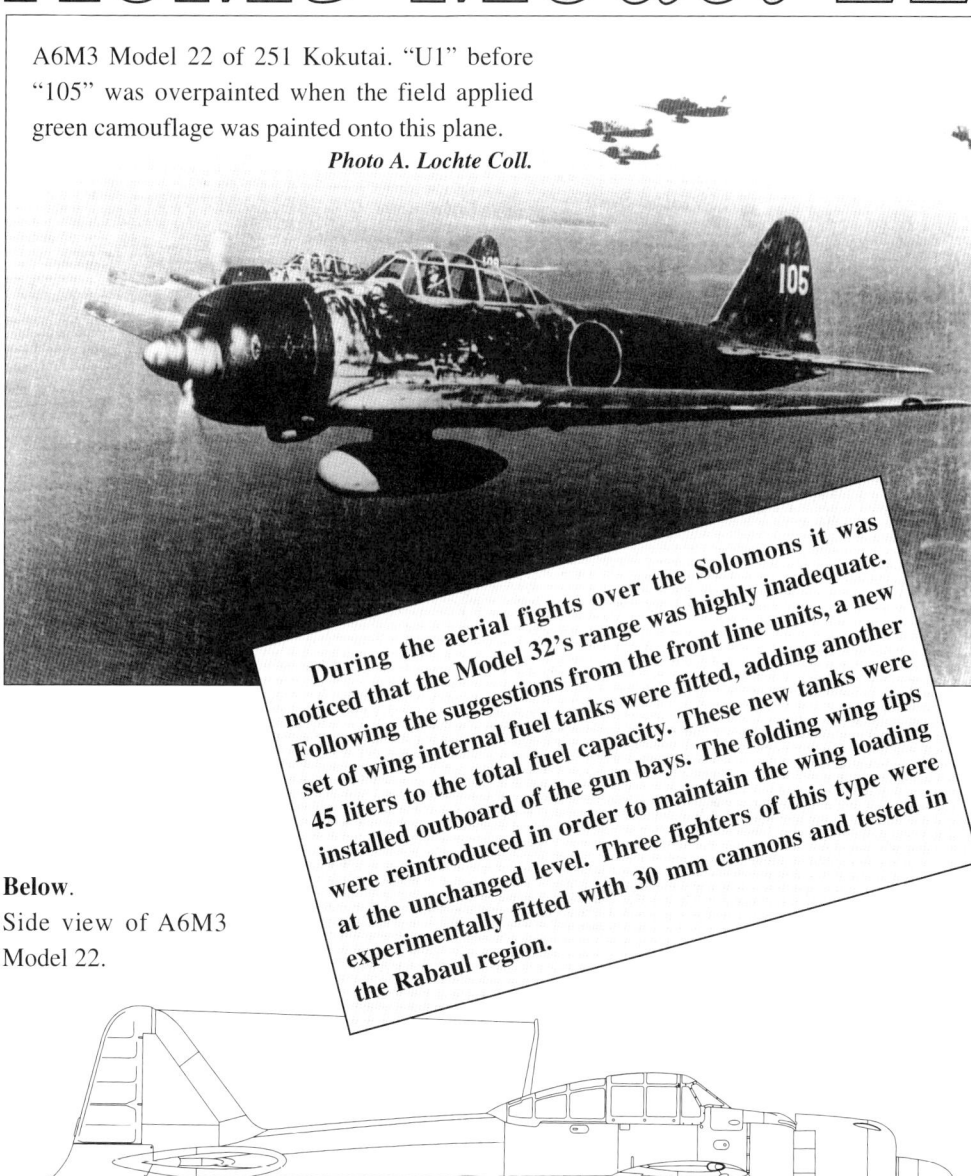

A6M3 Model 22 of 251 Kokutai. "U1" before "105" was overpainted when the field applied green camouflage was painted onto this plane.
Photo A. Lochte Coll.

During the aerial fights over the Solomons it was noticed that the Model 32's range was highly inadequate. Following the suggestions from the front line units, a new set of wing internal fuel tanks were fitted, adding another 45 liters to the total fuel capacity. These new tanks were installed outboard of the gun bays. The folding wing tips were reintroduced in order to maintain the wing loading at the unchanged level. Three fighters of this type were experimentally fitted with 30 mm cannons and tested in the Rabaul region.

Below.
Side view of A6M3 Model 22.

Mitsubishi A6M ZERO

VERSIONS

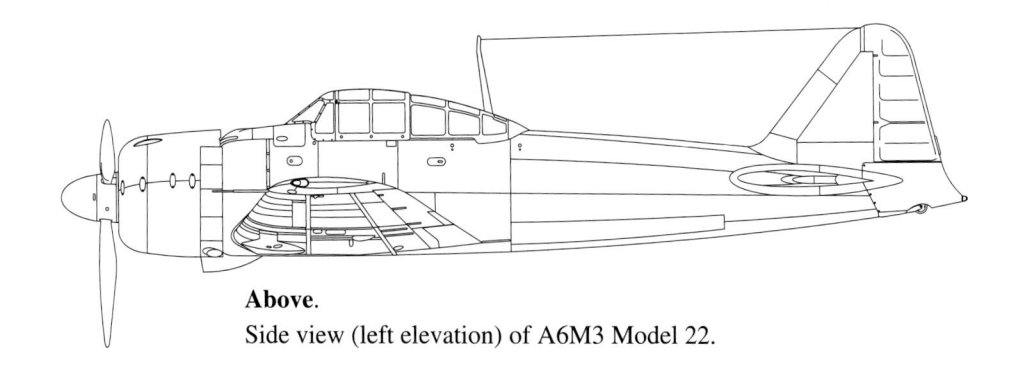

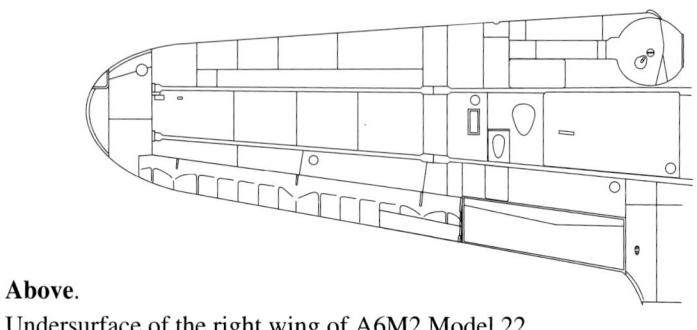

Above.
Side view (left elevation) of A6M3 Model 22.

Above.
Undersurface of the right wing of A6M2 Model 22.

A6M3 Model 22 of 582 Kokutai at Kahili Airfield at Buin on Bourgainville Island during operation I-Go in April 1943.

Photo. A. Lochte coll.

20 Mitsubishi A6M ZERO

VERSIONS

Mitsubishi A6M3 Model 22a

Beginning in December 1942 to June 1943 half of the A6M3 production were equipped with long barreled 20mm cannon Type 99 Mark 2 Model 3. This therefore included the last of the Model 32s. From July 1943 all of the A6M3s (all Model 22s) coming off the assembly line were so outfitted.

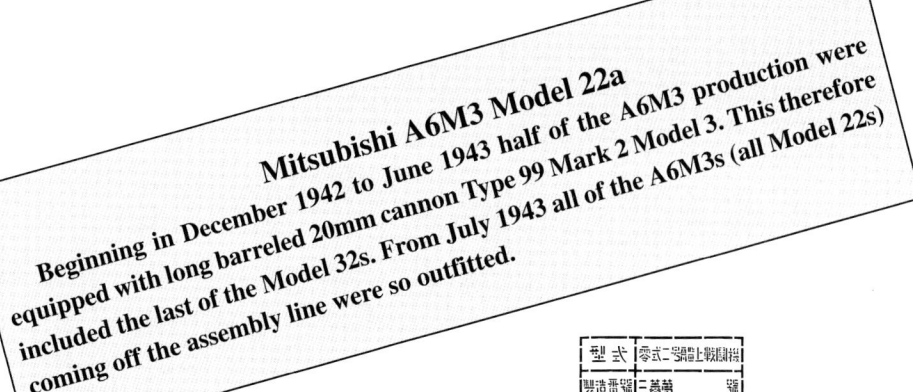

Stencil used on Model 22 built by Mitsubishi Factory.

Right.
Undersurface of the right wing of A6M2 Model 22a. Please note long barrel gun. (1)

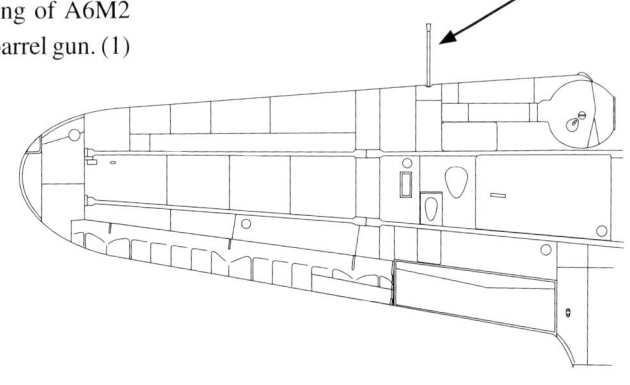

Stencil introduced on Model 22 built by Mitsubishi Factory in January 1943.

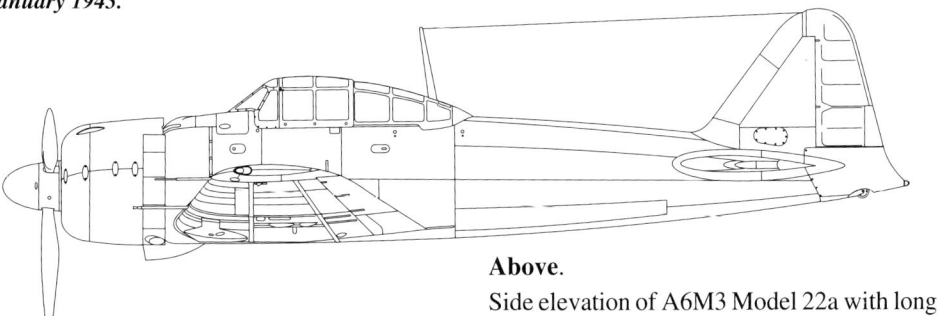

Above.
Side elevation of A6M3 Model 22a with long barrel cannons.

Mitsubishi A6M ZERO **21**

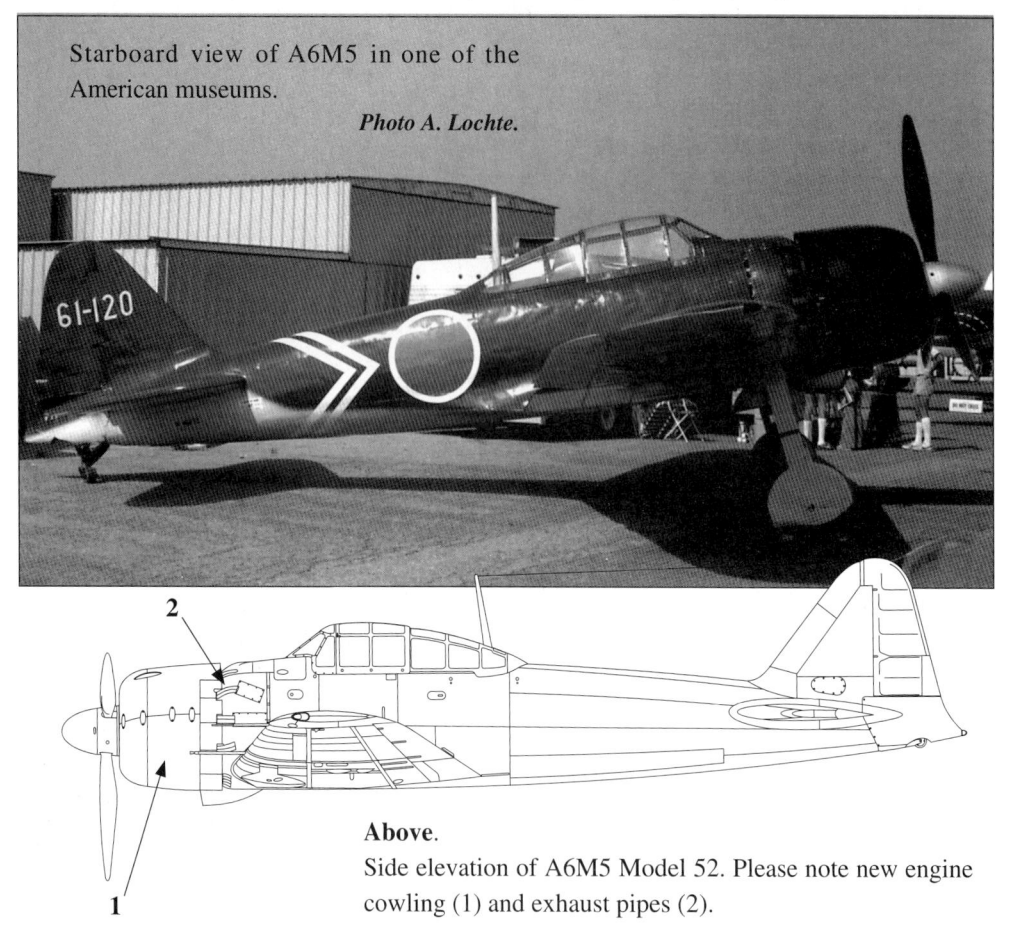

VERSIONS
A6M5 Model 52

Starboard view of A6M5 in one of the American museums.
Photo A. Lochte.

Above.
Side elevation of A6M5 Model 52. Please note new engine cowling (1) and exhaust pipes (2).

One of the primary weaknesses of the Zero's earlier versions was its insufficient dive speed. In order to improve this situation, a serial aircraft A6M3 Model 32 No. 904 was pulled straight from the assembly line and used as a test bed for improvements. Wing structure was reinforced, receiving a heavier gauge skin. Although the wingspan didn't change from the Model 32, the wingtips got a new, rounded-off shape. The engine remained the same with the exhaust collector ring replaced by individual exhaust stacks for each cylinder head. All of these modifications yielded a maximum dive speed of 660 kph. In August of 1942, this new model was accepted by the JIN, receiving the official designation as a Type 0 Carrier-Based Fighter Model 52.

VERSIONS

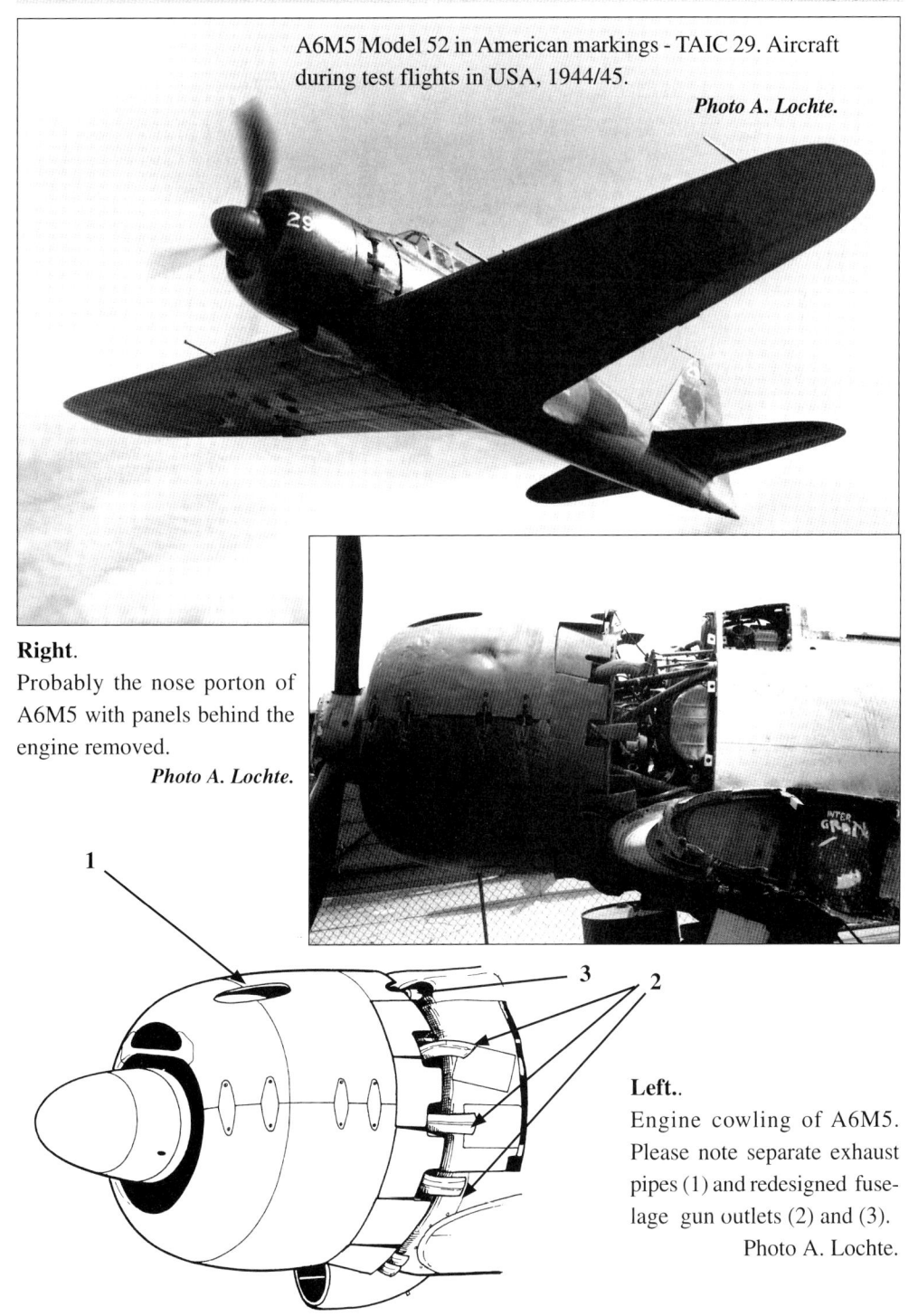

A6M5 Model 52 in American markings - TAIC 29. Aircraft during test flights in USA, 1944/45.
Photo A. Lochte.

Right.
Probably the nose porton of A6M5 with panels behind the engine removed.
Photo A. Lochte.

Left..
Engine cowling of A6M5. Please note separate exhaust pipes (1) and redesigned fuselage gun outlets (2) and (3).
Photo A. Lochte.

Mitsubishi A6M ZERO **23**

VERSIONS

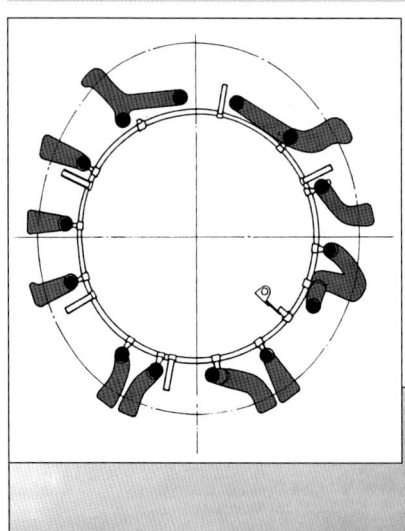

Left.
Drawing showing A6M5 exhaust pipes, individual exhaust stacks for each cylinder head

Stencil used on Model 52 built by Mitsubishi Factory.

Once again A6M5 Model 52 in American markings - TAIC 29.
Photo A. Lochte.

Above.
Side elevation of A6M5 Model 52 with details of arrestor hook.

24 *Mitsubishi A6M ZERO*

VERSIONS

Below.
Plan view of A6M5 Model 52.

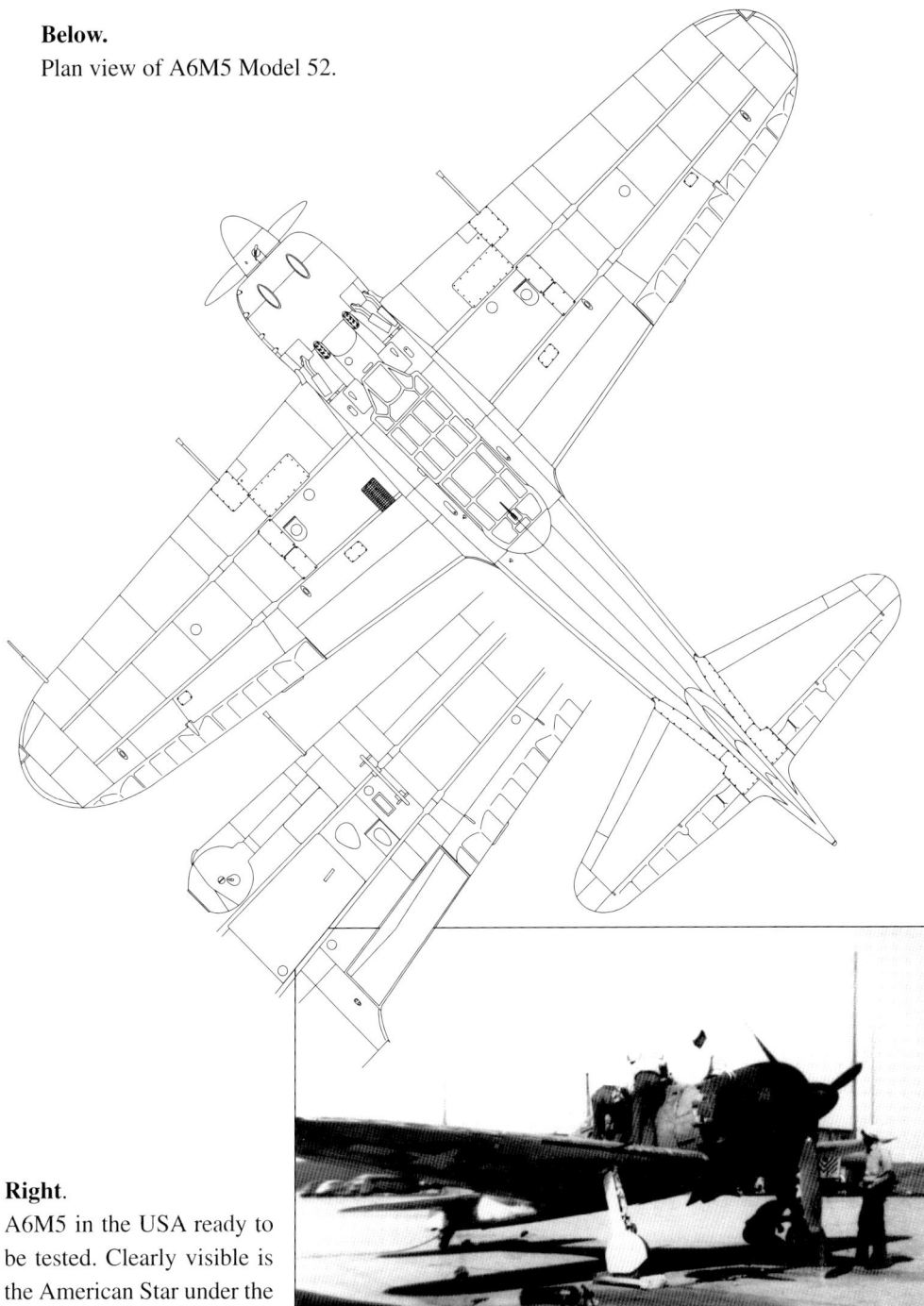

Right.
A6M5 in the USA ready to be tested. Clearly visible is the American Star under the right wing.
Photo A. Lochte Coll.

Mitsubishi A6M ZERO 25

VERSIONS

Once again an A6M5 preserved in American museum, but left side this time.
Photo A. Lochte.

Left.
A6M5 Model 52 in Kamikaze dive into flight deck of carrier USS Essex.

Photo A. Lochte Coll.

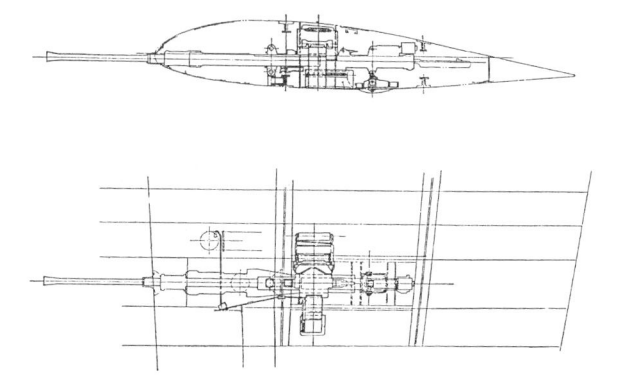

Right.
Drawing showing cannon 20 mm Type 99 Mark 2 Model 3, in the wing of A6M5.

26   *Mitsubishi A6M ZERO*

VERSIONS

Mitsubishi A6M5a Model 52a

By the beginnig of 1944 a new version of the Zero fighter was under way. With the wing covering of even heavier gauge, the maximum dive speed reached 740 kph. The armament went through some improvements too. Type 99 Model 1 cannons were replaced by new Type 99 Mark 2 Model 4s, fed byammunition belts, which also permitted the number of rounds to be increased to 125.

Below.
Wings of A6M5a. View of the left wing.

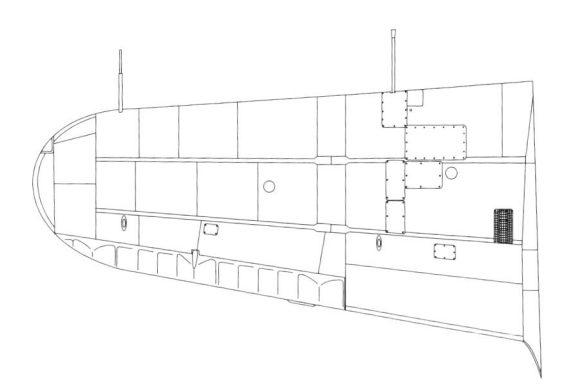

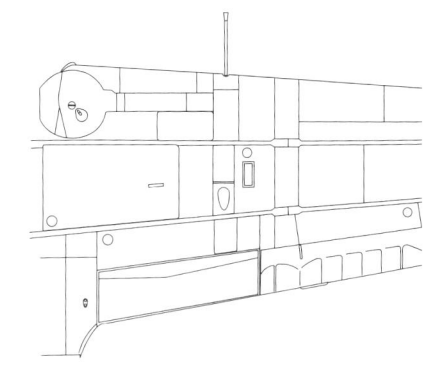

Left.
Bomb rack under the fuselage of the A6M5. Installation used on Zeros converted to fighter-bomber.

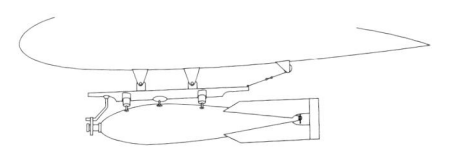

Left.
Zero in sight. Gun camera pictures from an American aircraft.
Photo A. Lochte Coll.

Stencil used on Model 52a.

Mitsubishi A6M ZERO **27**

VERSIONS

Mitsubishi A6M5b Model 52b

Concurrently with the A6M5a work continued on the A6M5b version. With this version, however, the main emphasis was on appropriate protection for the pilot as well as the most vulnerable elements of the aircraft itself. Fuel tanks received a new, CO_2 based fire extinguishing system, while the front part of the windshield was reinforced by installing a 50 mm thick armoured glass. The total firepower was improved upon by replacing one of the fuselage-mounted 7.7 mm machine guns with a more powerful 13.2 mm Type 3 gun.

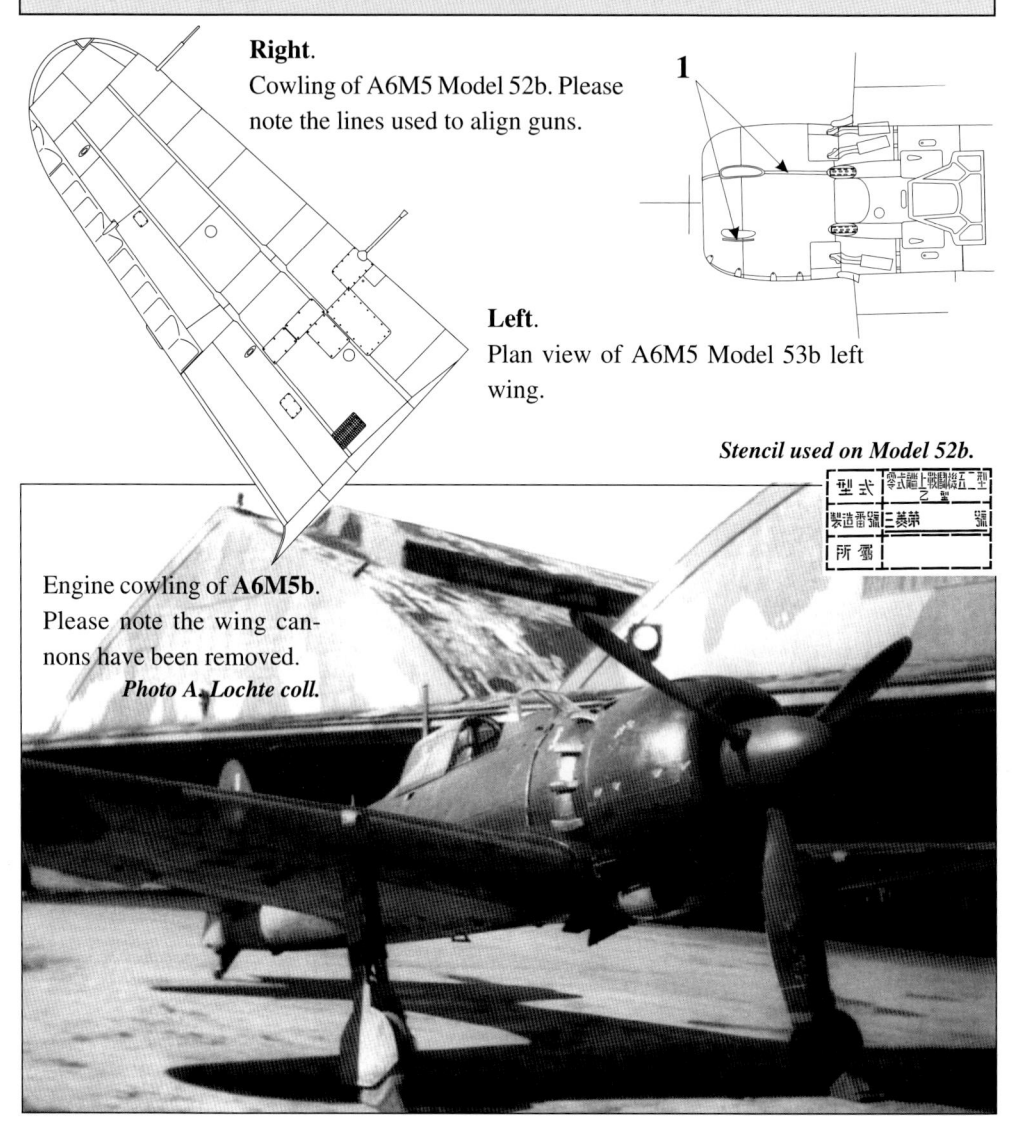

Right.
Cowling of A6M5 Model 52b. Please note the lines used to align guns.

Left.
Plan view of A6M5 Model 53b left wing.

Stencil used on Model 52b.

Engine cowling of **A6M5b**. Please note the wing cannons have been removed.
Photo A. Lochte coll.

28   *Mitsubishi A6M ZERO*

Mitsubishi A6M5c Model 52c

In April 1944 the Imperial Japanese Navy placed an order for a modified version of the Zero fighter. In response to the IJN specifications, the A6M5c Model 52c was introduced in October. The pilot protection was additionally improved by installation of an armoured plate immediately behind the pilot's seat. In order to increase the range, the new model was fitted with a 70 liter self-sealing fuel tank, located behind the cockpit. In addition, the armament was modified adding two wing-mounted 13.2mm machine guns, Type 3. The fuselage-mounted 7.7mm gun was deleted, leaving just a single 13.2mm gun. Wings were provided with racks for air-to-air rocket projectiles. In order to compensate for the increased weight, a new and more powerful engine was called for. The Navy chose the Sakae 31A, with water-methanol injection for short bursts of extra power. However, the new engine was still under development, and suffered from serious problems requiring more tests. This meant that production aircraft retained the original engine - the Sakae 21 - and was thus seriously underpowered. The debut of this new model was catastrophic, virtually all the A6M5c aircraft in action being shot down by US fighters.

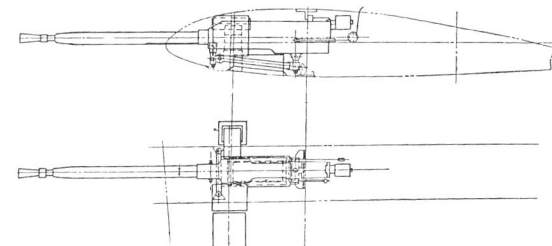

Left.
Machine-guns, 13 mm installation in the wing of A6M5c.

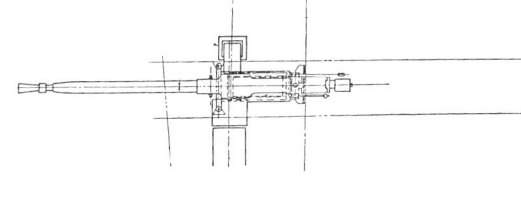

Right.
Plan view of A6M5c Model 52c wing. Please note cannon and machine gun barrels and rocket projectile mounts (1).

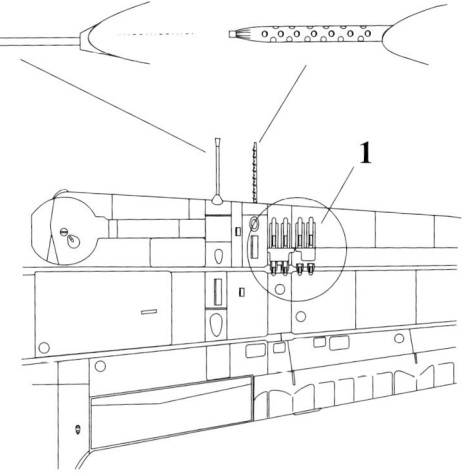

VERSIONS

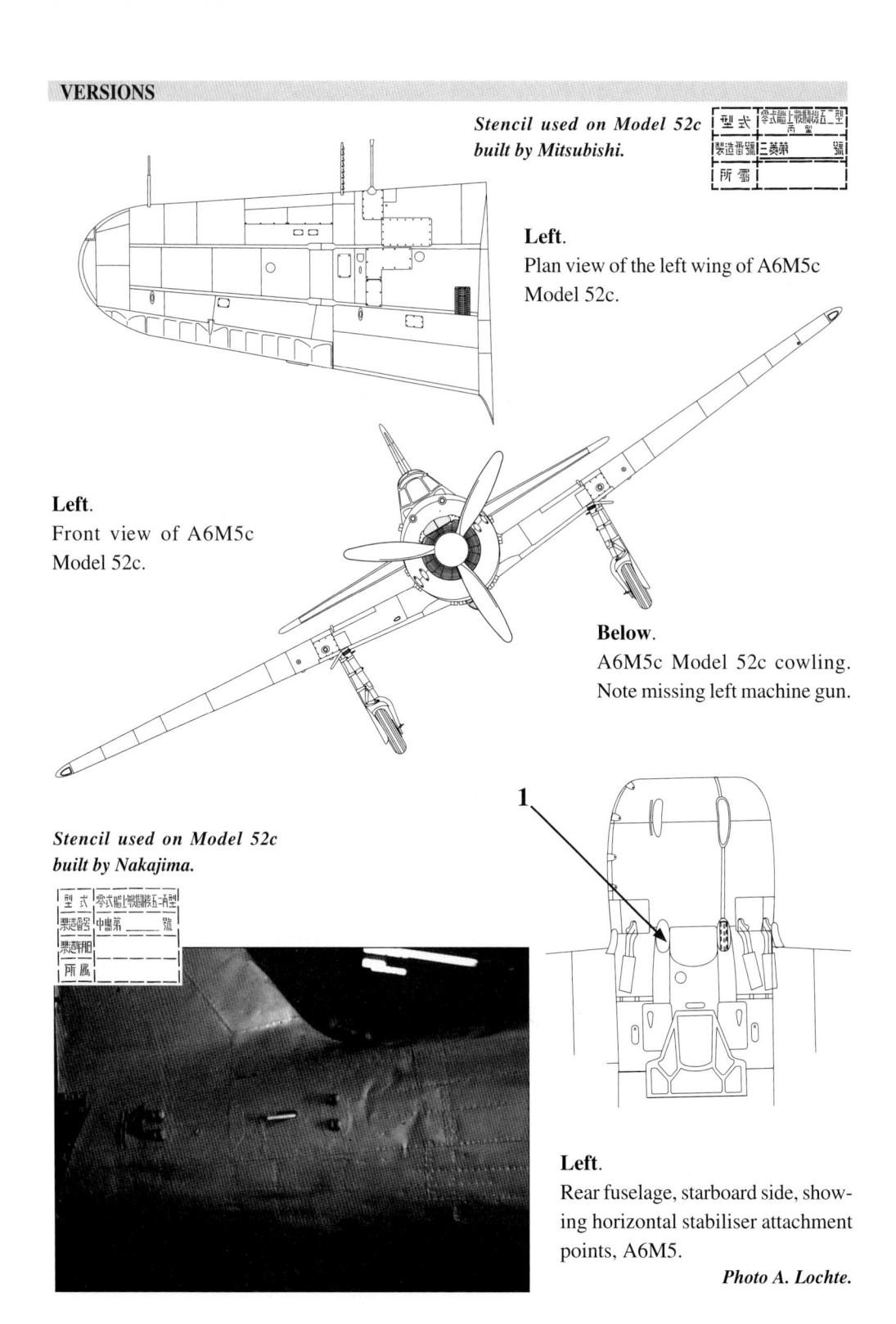

Stencil used on Model 52c built by Mitsubishi.

Left.
Plan view of the left wing of A6M5c Model 52c.

Left.
Front view of A6M5c Model 52c.

Below.
A6M5c Model 52c cowling. Note missing left machine gun.

Stencil used on Model 52c built by Nakajima.

Left.
Rear fuselage, starboard side, showing horizontal stabiliser attachment points, A6M5.

Photo A. Lochte.

30   Mitsubishi A6M ZERO

VERSIONS

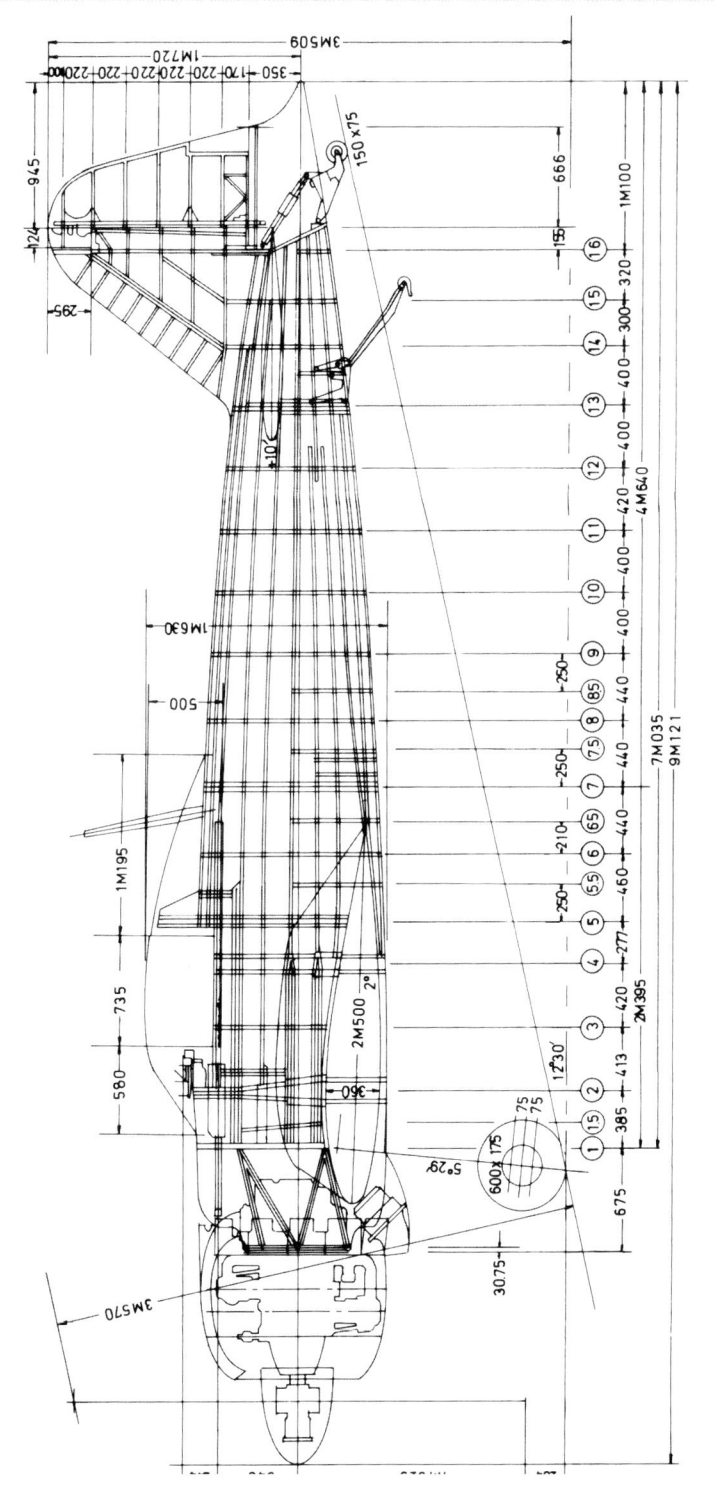

Fuselage dimensions of A6M5 Model 52.
Drawing from original Technical Manual.

Mitsubishi A6M ZERO 31

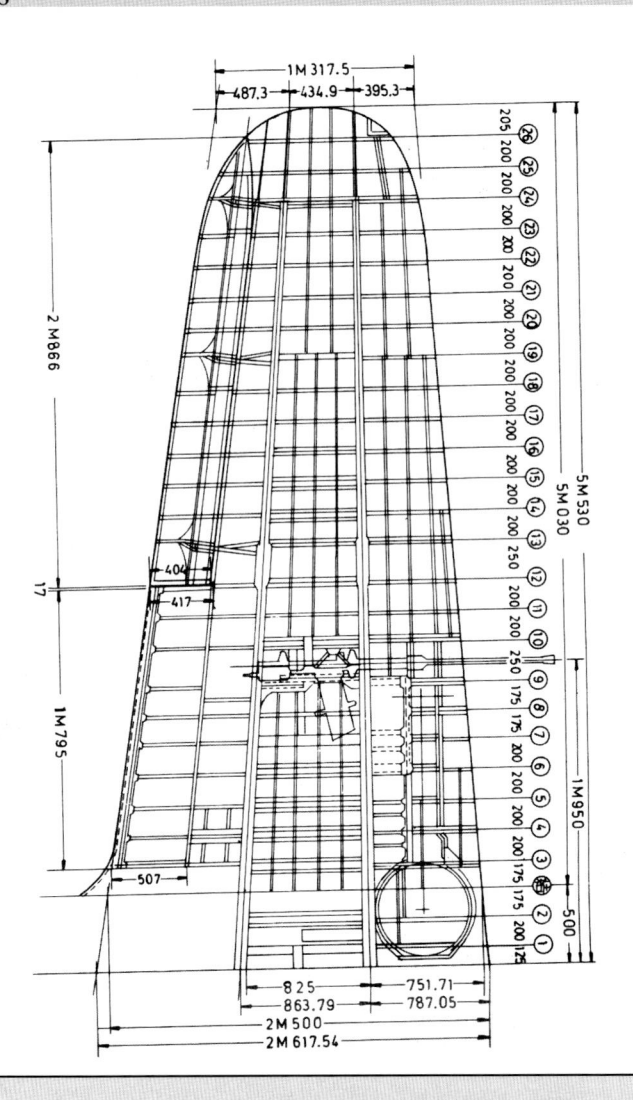

VERSIONS

Wing dimensions of A6M5 Model 52. Drawing from original Technical Manual.

Mitsubishi A6M6c model 53c

In November 1944, the new engine, the Sakae 31A, finally became available after the completion of tests, and a serial A6M5c was equipped with it at the Mitsubishi plant. As a result, the maximum speed in level flight increased to 555 kph. Serial aircraft also received new, self-sealing fuel tanks in the wings. Nonetheless, the only one aircraft produced at this phase of the war were plagued with numerous defects (both power plant and the aircraft). The troublesome fuel injection system was a major problem.

Mitsubishi A6M ZERO

VERSIONS
A6M7 Model 63

In mid 1944, with the loss of all of the larger aircraft carriers, the IJN high command requested yet another modification of the Zero. This time it was to have a role of dive bomber. Its new incarnation required the ability to dive with great speeds, thus necessitating additional reinforcements to the covering of the fuselage rear parts and wings. Under-fuselage drop tank installation was replaced with a Mitsubishi designed bomb rack, accommodating a 250 kg bomb. Provision were also made for two 350 liters under-wing fuel tanks, attached outboard of the wheel wells. It is not known how many machines of this type were built.

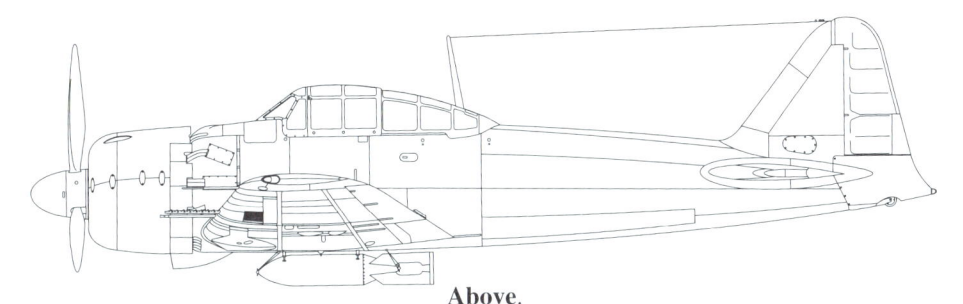

Above.
Side view of A6M7 Model 63. Aircraft shown with bomb on the bomb rack under fuselahge.

Right.
Engine and spinner of A6M7.
Photo A. Lochte.

Stencil used on Model 63 built by Mitsubishi.

VERSIONS

A6M8 Model 64

Side view of A6M8 Model 64. Aircraft with removed fuselage armament (2), redesigned engine cowling (1) and windscreen of new shape (3).

Disappointing performance of models A6M5c and A6M7 resulted in replacing the previously used Sakae engines with a new power plant - the Kinsei 62. Again, the new engine installation required changes to the fuselage's front portion, and among other things the only fuselage machine gun had to be removed, leaving the aircraft with wing guns only. Similarly to the previous version, the underwing drop tanks were used while the fuselage centerline bomb rack could now accommodate a 500 kg bomb. An aircraft such equipped, was capable of reaching a maximum speed of 577 kph. In May of 1945 an order for 6500 aircraft was officially placed, receiving the highest priority in the armament industry. The serial production of this model, however, never got under way.

Right.
A6M7 Model 63 in a museum. Aircraft beautifully restored and preserved in excellent condition.
Photo A. Lochte.

COLOUR PROFILES

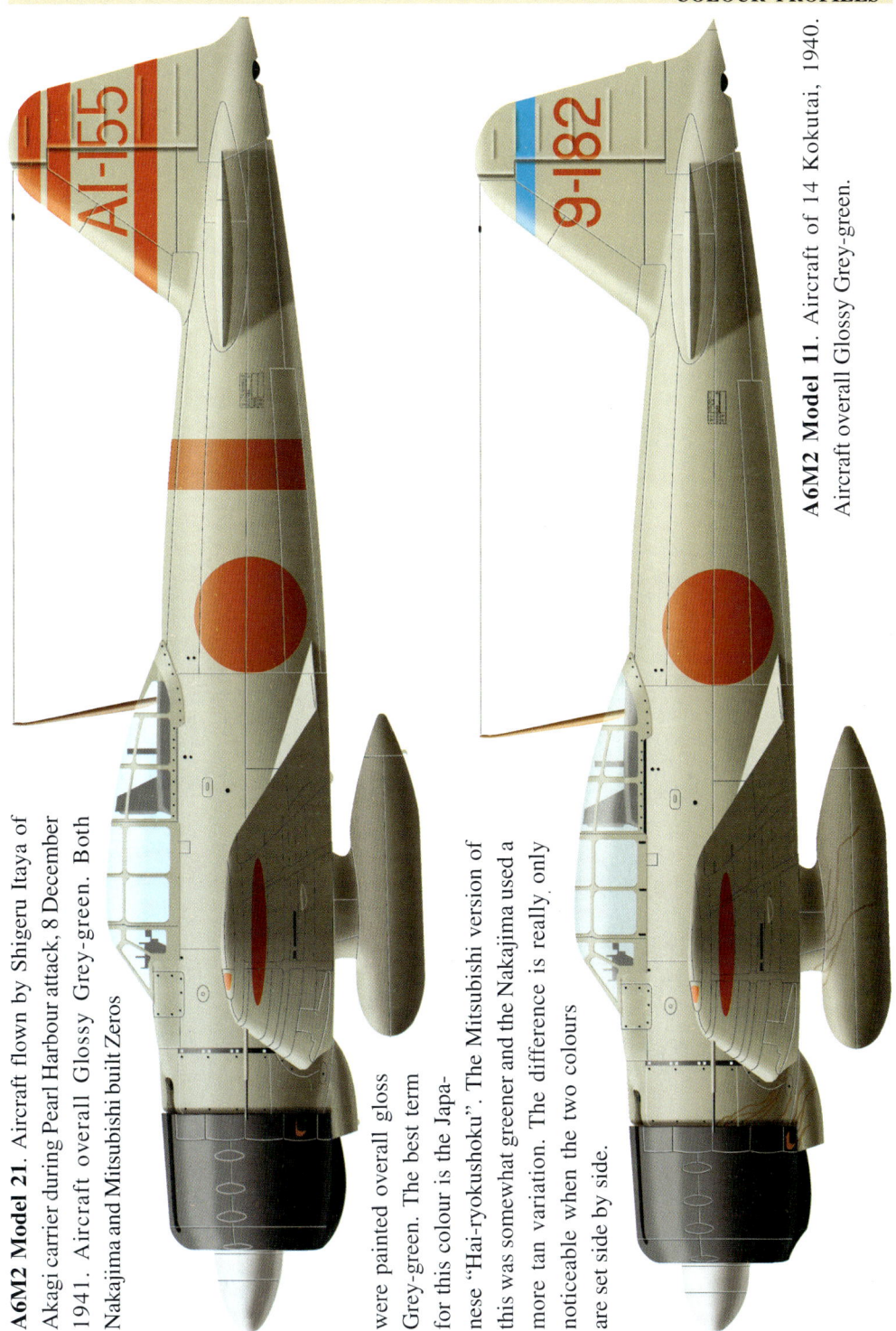

A6M2 Model 21. Aircraft flown by Shigeru Itaya of Akagi carrier during Pearl Harbour attack, 8 December 1941. Aircraft overall Glossy Grey-green. Both Nakajima and Mitsubishi built Zeros were painted overall gloss Grey-green. The best term for this colour is the Japanese "Hai-ryokushoku". The Mitsubishi version of this was somewhat greener and the Nakajima used a more tan variation. The difference is really only noticeable when the two colours are set side by side.

A6M2 Model 11. Aircraft of 14 Kokutai, 1940. Aircraft overall Glossy Grey-green.

Mitsubishi A6M ZERO **35**

COLOUR PROFILES

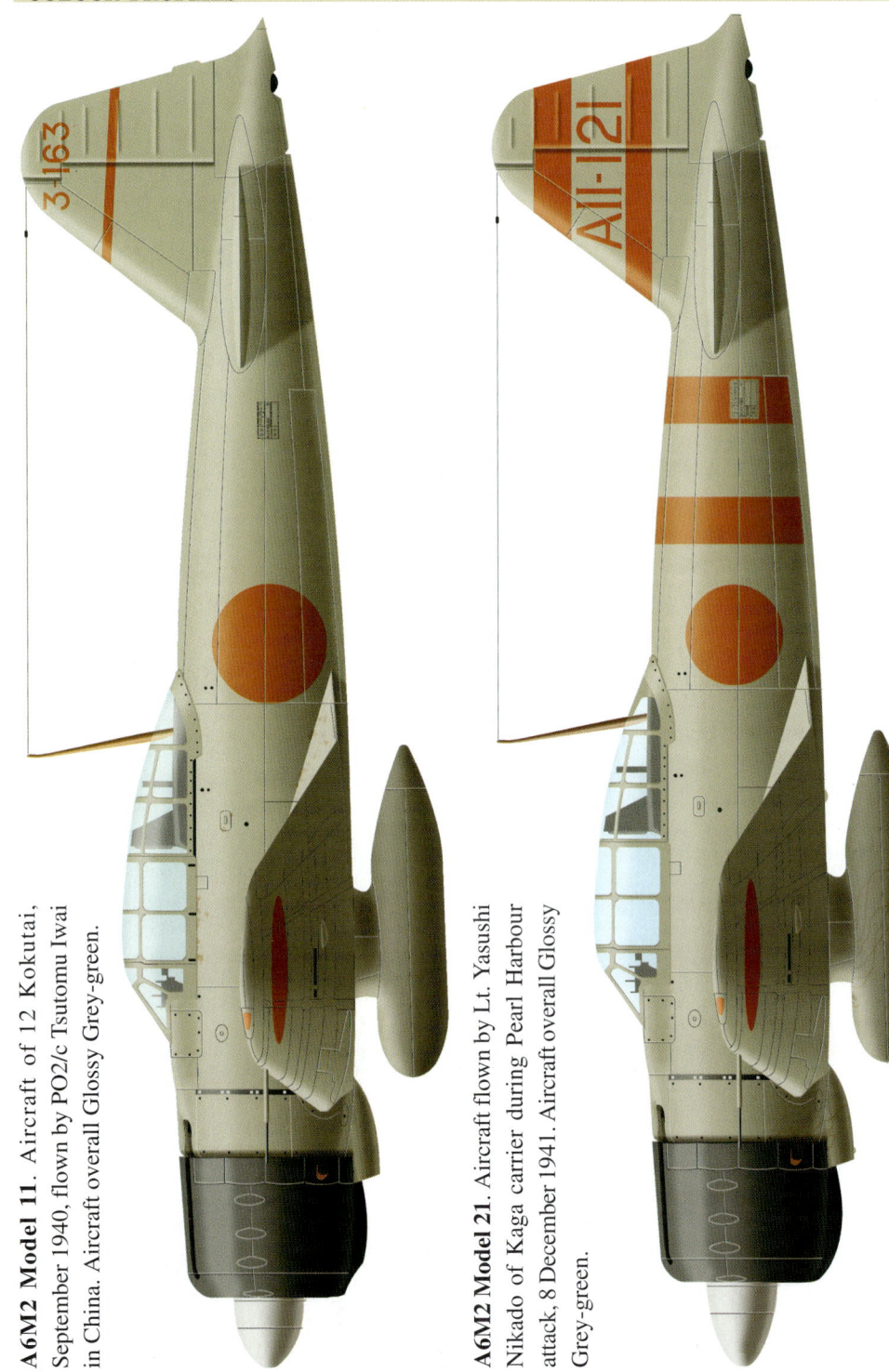

A6M2 Model 11. Aircraft of 12 Kokutai, September 1940, flown by PO2/c Tsutomu Iwai in China. Aircraft overall Glossy Grey-green.

A6M2 Model 21. Aircraft flown by Lt. Yasushi Nikado of Kaga carrier during Pearl Harbour attack, 8 December 1941. Aircraft overall Glossy Grey-green.

DETAILS

Wing

A6M5 restored to flying condition.
Photo S. T. Hards via J. Kightly

Light.
Wing tip of A6M5.
Photo A. Lochte.

Right.
Folded wing tip of A6M3 Model 22. This same feature was used on Model 21.
Photo A. Lochte

Mitsubishi A6M ZERO 37

DETAILS

Right.
Folded wing tip of A6M2 replica. Construction is similar to the A6M3 Model 22, especially when compared to photo on the pervious page.
Photo R. Pęczkowski.

Left.
Details of the folded wing tip, A6M3 Model 22.
Photo A. Lochte

Left.
Port wing tip - hinge line and handle of A6M3 model 22.
Wing tip in flying position.
Photo A. Lochte

38   *Mitsubishi A6M ZERO*

DETAILS

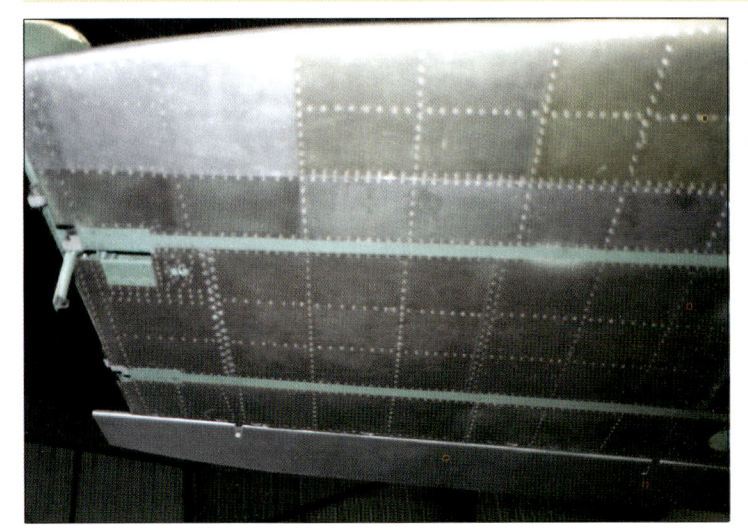

Starboard wing tip in the folded position, A6M3 Model 22. Aircraft under restoration.
Photo A. Lochte.

Starboard wing of the A6M3 Model 22, during restoration. Inspection panels shown open.
Photo A. Lochte.

Wing tip and aileron control horn, starboard wing, A6M7.
Photo A. Lochte.

DETAILS

Above.
Starboard flap, inner view, A6M7. Please note details of the flap construction

Photo A. Lochte.

Left.
Drawing from Technical Manual of A6M. Internal construction of the flap.

Right.
Central part of the starboard wing, upper surface of the flap area. Visible red line indicates the "no-step" area. A6M5.

Photo A. Lochte.

40   Mitsubishi A6M ZERO

DETAILS

Left.
Aileron control horn, starboard wing, A6M7.
Photo A. Lochte

Below.
Pitot tube, port wing. A6M7.
Photo A. Lochte.

Below.
Starboard wing of the A6M7.
Photo A. Lochte.

Mitsubishi A6M ZERO **41**

COLOUR PROFILES

A6M2 Model 21. Aircraft flown by Sumio Nouno of Hiryu carrier during Pearl Harbour attack, 8 December 1941. Aircraft overall Glossy Grey-green.

A6M2 Model 21. RAF Station Seletar, Singapore 1945. "ATAIU" - Allied Technical Air Intelligence Unit. "SEA" - South East Asia. Aircraft in British camouflage.

42   Mitsubishi A6M ZERO

COLOUR PROFILES

A6M2 Model 21 of Tainan Kokutai (251 Kokutai), 1942, flown by Saburo Sakai. Aircraft overall Glossy Grey-green.

A6M2 Model 21 from Tsukuba Kokutai, 1944. Upper surfaces - Dark green (Japanese - an-ryokushoku), under surfaces Glossy Grey-green. Dark green colour was painted over Grey-green.

Mitsubishi A6M ZERO   **43**

DETAILS

Green, starboard wing tip light of A6M7.
    Photo A. Lochte.

Underside of the wings at fuselage, showing wheel well, A6M5.
    Photo A. Lochte.

Underside of the port wing, A6M3.
    Photo A. Lochte.

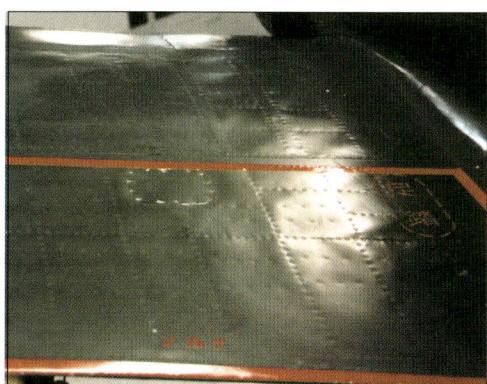

DETAILS

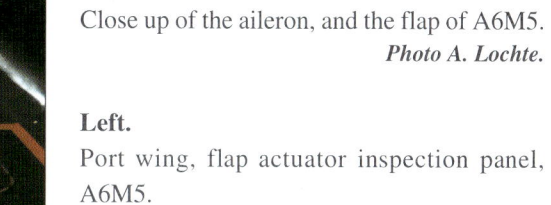

Above.
Close up of the aileron, and the flap of A6M5.
Photo A. Lochte.

Left.
Port wing, flap actuator inspection panel, A6M5.
Photo A. Lochte.

Left.
Port aileron control horn, mass balance is missing, A6M3.
Photo A. Lochte.

DETAILS

Right.
Cockpit air intake, port wing, A6M3. Please note that on early Model 11 air intake was elliptical in shape.
Photo A. Lochte.

Right, below.
Two photos of landing gear position indicator, A6M3.
Photos A. Lochte.

Below.
Starboard wing, showing flap actuator inspection panel, fuel filler cover and navigation light, A6M3.
Photo A. Lochte.

46   *Mitsubishi A6M ZERO*

DETAILS

Left.
The bottom port mid wing section, showing cannon blisters and empty shell chute, as viewed from the wing tip, A6M5.
Photo A. Lochte.

Below.
Pitot tube, port wing, A6M7.
Photo A. Lochte.

Above.
Pitot tube, earlier type, port wing, A6M3.
Photo A. Lochte.

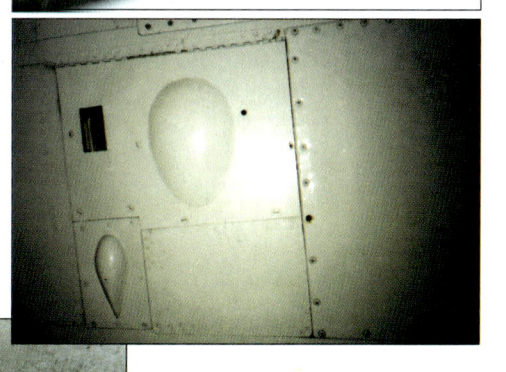

Above.
The bottom port mid wing section, showing cannon blisters and empty shell chute, A6M5.
Photo A. Lochte.

Left.
Underneath of the fuselage at wing, A6M3.
Photo A. Lochte.

Mitsubishi A6M ZERO   47

DETAILS

Right.
Underside of the port wing, A6M3.
Photo A. Lochte.

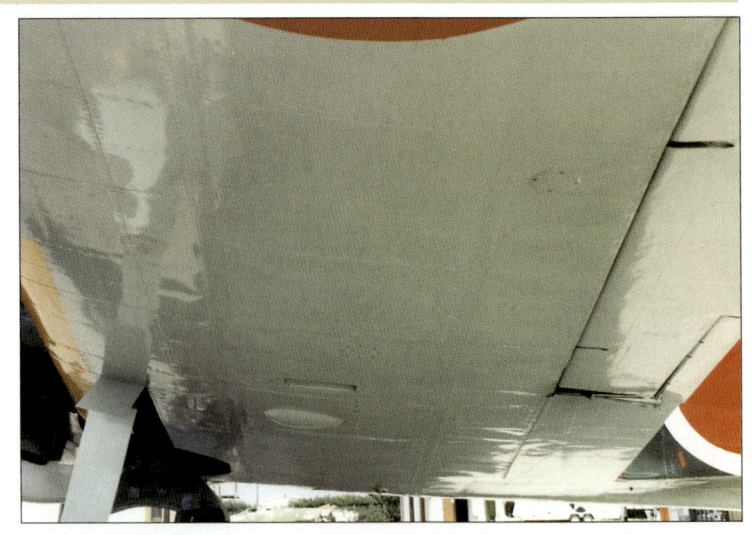

Right.
Red, port wing tip light. A6M5.

Below.
Navigation light on the starboard wing, A6M3.

Below, right.
Port mid wing section showing faired over housing of the bomb rack, A6M7.
All photos A. Lochte.

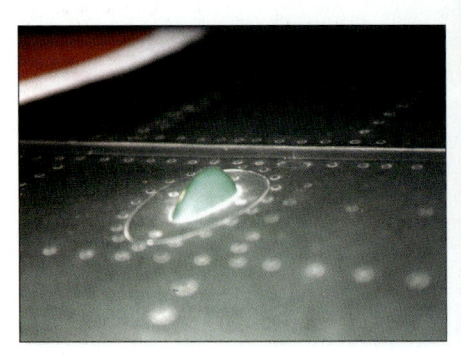

48   *Mitsubishi A6M ZERO*

DETAILS

Fuselage

Left.
A6M5 Model 52c in museum. Aircraft preserved in very good condition.
Photo A. Lochte.

Left and below.
Beautifully restored A6M2 in the Australian War Memorial Museum, Canberra. Aircraft in the camouflage as flown by Saburo Sakai. Please note that according to the latest research colour of early Zero Models 21 and Model 32 was different from what is shown in this picture. See colour profiles.
Photos R. Wallsgrove.

Mitsubishi A6M ZERO

DETAILS

Right.
Oil cooler intake, underside view, A6M5. Wheel well covers are also visible.
Photo A. Lochte.

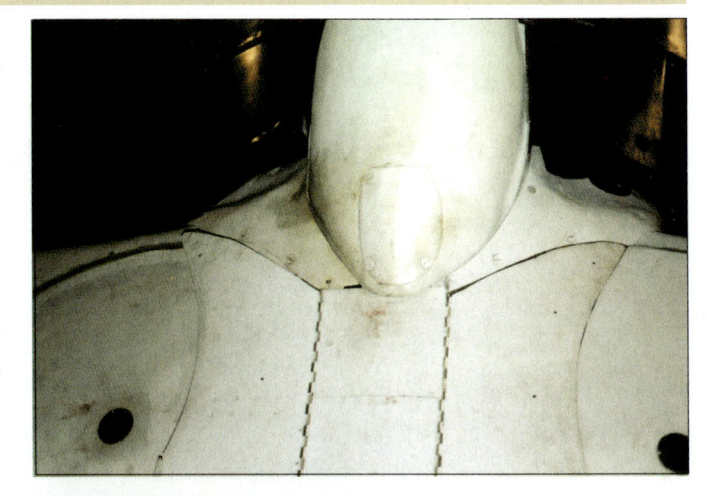

Middle.
Retractable step and inspection panel. Port side beneath canopy, A6M5.
Photo A. Lochte.

Right, below.
Port side upper fuselage between cowl flaps and windscreen, A6M5.
Photo A. Lochte.

Below.
The wing fillet at fuselage.
Photo A. Lochte.

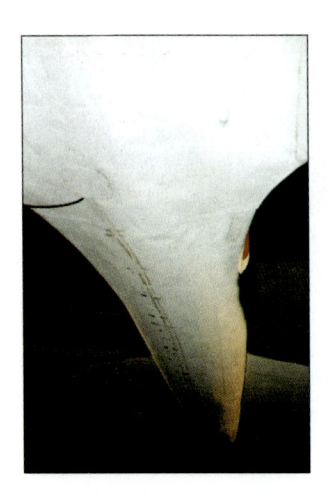

50   Mitsubishi A6M ZERO

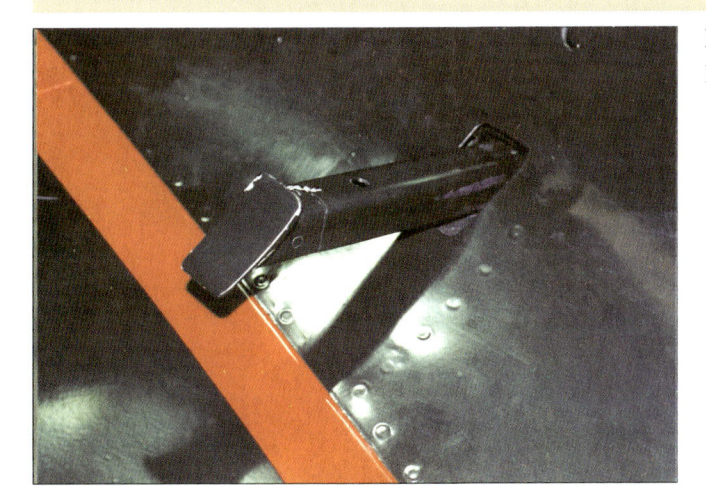

DETAILS

Left.
Retractable step, A6M3.
Photo A. Lochte.

Left.
Retractable step, just above the wing fillet.
Photo A. Lochte.

Left.
Middle part of the fuselage, A6M7.
Photo A. Lochte.

Mitsubishi A6M ZERO **51**

COLOUR PROFILES

A6M2 Model 21 of 301 Hikotai, 201 Kokutai flown by Yukio Seki during one of the first kamikaze attacks, October 1944. Upper surfaces Dark Green, under surfaces Grey-green. Please note that in the case of Mitsubishi at least, the Grey green paint was applied over the entire aircraft and then the dark green painted over that. Thus the dark green would chip to show the Grey-green paint under it first. As the planes were supposedly carrier based they were well primed before the final coat of paint was applied. Any paint chips would tend to show red brown primer instead of the aluminium skin.

A6M2 Model 21 of 203 Kokutai, 1944. Upper surfaces Dark Green, under surfaces Grey-green.

52 *Mitsubishi A6M ZERO*

COLOUR PROFILES

A6M2 Model 21 of Ohita Kokutai in 1944. Upper surfaces Dark Green, under surfaces in orange (tou-ou-shoku).

A6M3 Model 22 of Iwakumi Kokutai in 1943. Aircraft overall Glossy Grey-green.

Mitsubishi A6M ZERO **53**

DETAILS

Above. Fuselage of A6M3 during restoration. Oil tank on firewall is visible.

Photo A. Lochte

Below. Internal fuselage structure of Zero A6M2. Sadly it is only a replica, built in Russia.

Photo R. Pęczkowski.

DETAILS

Left.
Once again internal fuselage structure of Zero A6M2. Close up view of the tail structure.
Photo R. Pęczkowski.

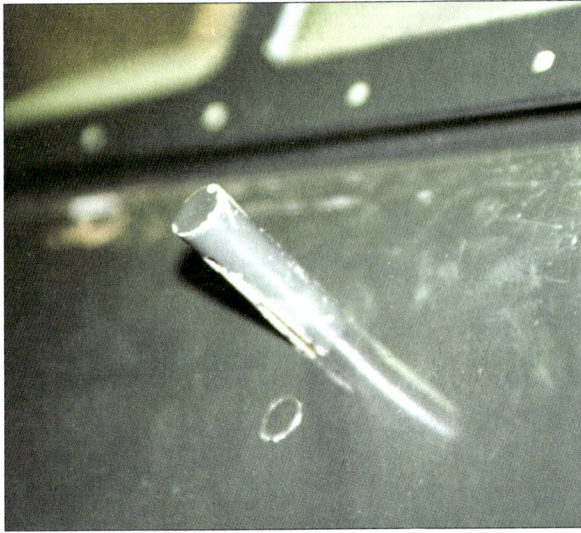

Left.
A6M3. The hand grip helpful when entering cockpit.
Photo A. Lochte.

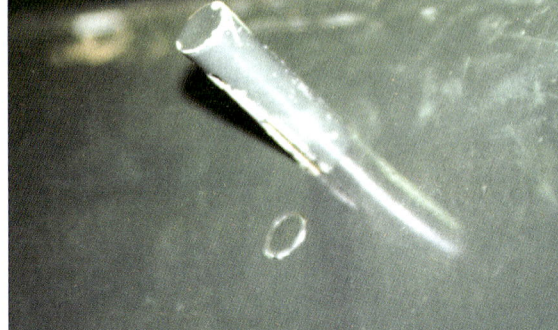

Left, below.
Remains of a Zero in the IWM, Duxford. Please note colour of the primer enamel - almost brown.
Photo R. Pęczkowski

Below.
Oil cooler intake. A6M3.
Photo A. Lochte.

Mitsubishi A6M ZERO 55

DETAILS

Right.
Side view of the A6M2 remains in Duxford.
Photo R. Pęczkowski

Right.
Cover removed showing compressed air filler. Port fuselage side below rear portion of the canopy.
Photo A. Lochte.

Below.
Front view of A6M3 Model 22.
Photo A. Lochte.

56 *Mitsubishi A6M ZERO*

DETAILS

Left and middle of the page.
Remains of A6M2 Model 21. See also photos on page 5 and 10.

Photos R. Pęczkowski

Bottom of the page.
The same aircraft as in the photo on opposite page, but side view with panels behind the engine removed.

Photo A. Lochte.

Mitsubishi A6M ZERO 57

DETAILS

Cockpit & Canopy

Right.
Headrest - port side of the A6M3.
Photo A. Lochte.

Right below.
Rear of the canopy of A6M3.
Photo A. Lochte.

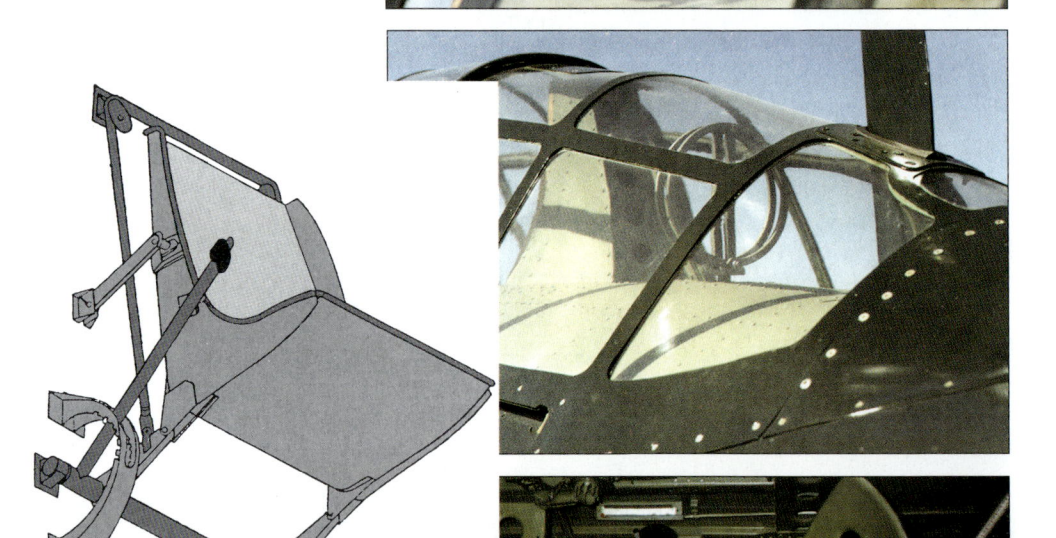

Above.
Drawing of the pilot's seat. Drawing from original Technical Manual.

Right.
Pilot's seat A6M2 Model 21.
Photo R. Wallsgrove.

58   *Mitsubishi A6M ZERO*

DETAILS

Left.
Cockpit of A6M2 Model 21. Well visible is instrument panel, port console and control column.
Photo R. Wallsgrove.

Left.
Instrument panel of A6M2 Model 21. Please note how machine guns were mounted in the cockpit.
Photo R. Wallsgrove.

Mitsubishi A6M ZERO **59**

DETAILS

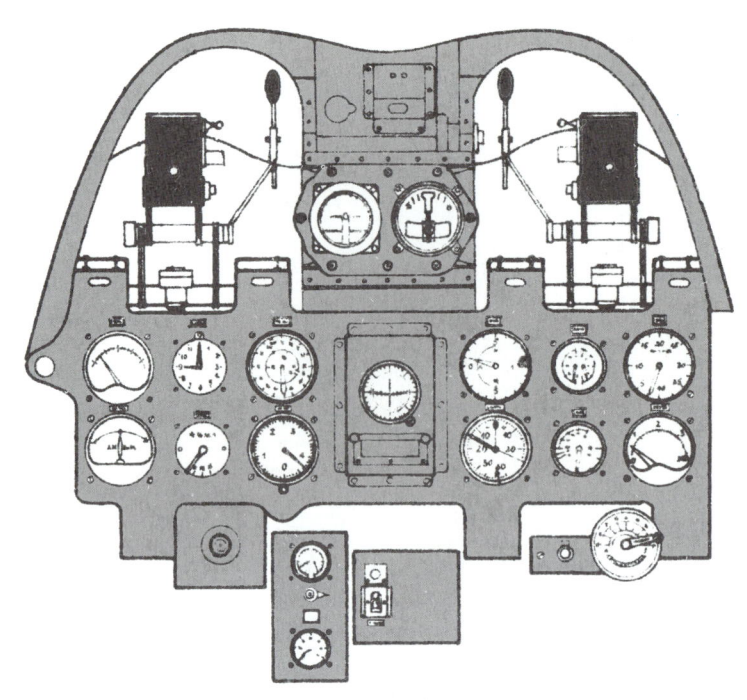

Above. Drawing of instrument panel of A6M5.

Below. Canopy of A6M5 showing the wooden grips of the machine-gun charging cocks.
Photo A. Lochte.

DETAILS

Left.
Headrest and canopy lock handle of A6M3.
Photo A. Lochte.

Left.
Windscreen and the gun sight of A6M3.
Photo A. Lochte.

Left.
45 mm armoured glass used on A6M7.
Photo A. Lochte.

DETAILS

Right.
A6M7 windshield - upper view.
Photo A. Lochte

Below.
Canopy of A6M7.
Photo A. Lochte.

Below.
Side view of the cockpit canopy of A6M5. Canopy in closed position.
Photo A. Lochte.

62   *Mitsubishi A6M ZERO*

DETAILS

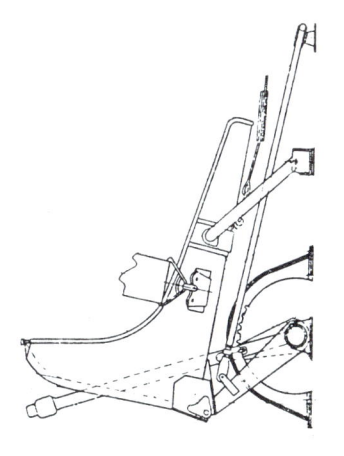

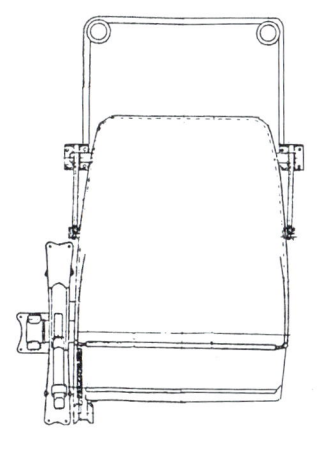

Left.
Pilot seat used till 1945.
Side and front view.

Left.
Headrest of A6M5.
Photo A. Lochte.

Left.
Side view of the windshield, A6M5.
Photo A. Lochte.

Mitsubishi A6M ZERO

DETAILS

Right.
Once again side view of the cockpit canopy, A6M5.
Photo A. Lochte.

Middle of the page.
Close up view of the A6M5 canopy.
Photo A. Lochte.

Bottom.
Left side of the canopy, close up view.
Photo A. Lochte.

64 *Mitsubishi A6M ZERO*

COLOUR PROFILES

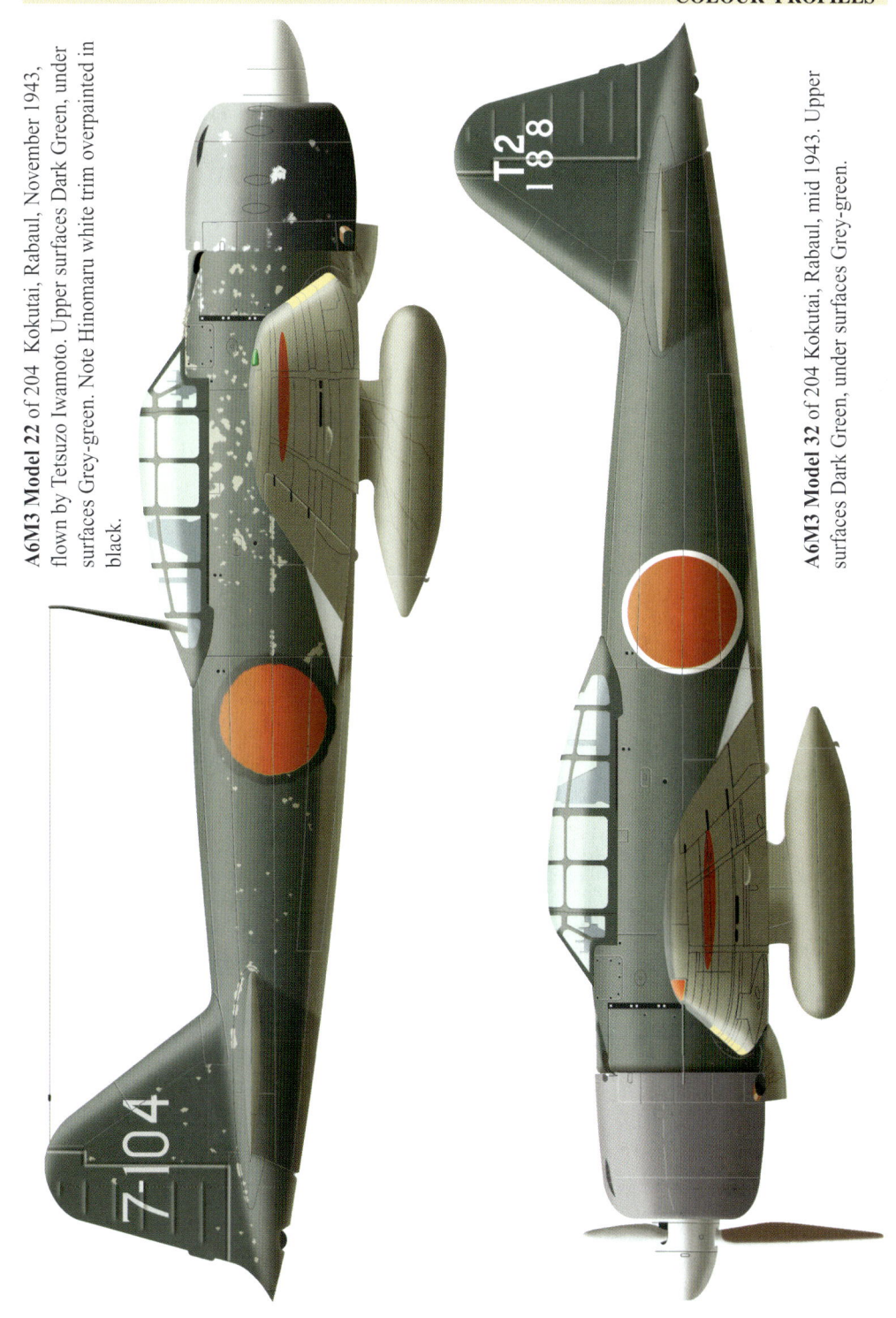

A6M3 Model 22 of 204 Kokutai, Rabaul, November 1943, flown by Tetsuzo Iwamoto. Upper surfaces Dark Green, under surfaces Grey-green. Note Hinomaru white trim overpainted in black.

A6M3 Model 32 of 204 Kokutai, Rabaul, mid 1943. Upper surfaces Dark Green, under surfaces Grey-green.

Mitsubishi A6M ZERO **65**

DETAILS

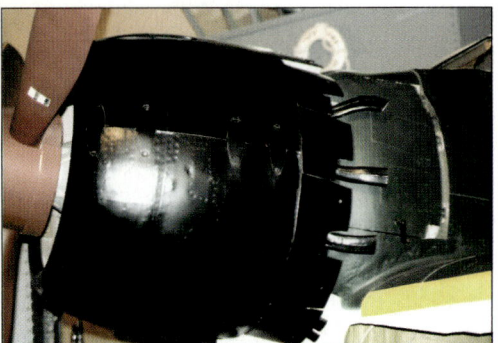

Engine

Three photos showing nose of A6M7. Spinner, individual exhaust pipes and details of the front part of the engine are visible.

Photos A. Lochte.

66 *Mitsubishi A6M ZERO*

DETAILS

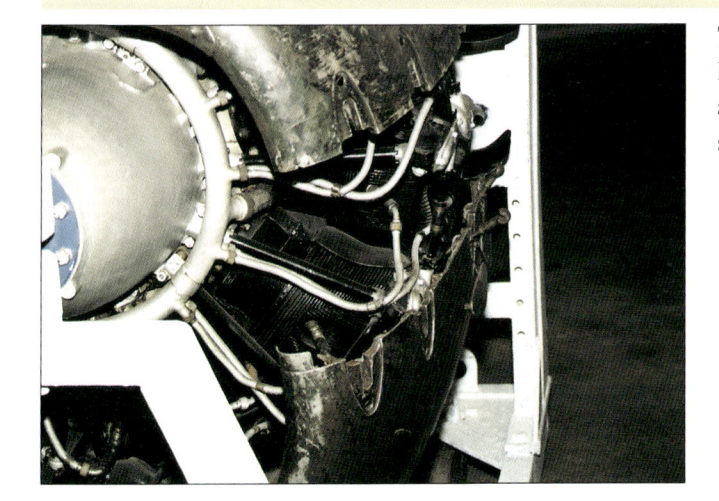

This page.
Photos of Sakae 21 engine and its cowling. Engine preserved at Cosford Museum.
Photos R Pęczkowski.

DETAILS

Above.
Details of Sakae engine on A6M2, during maintaince.

Photo R. Pęczkowski coll.

Left and opposite page left.
Panel behind cowling removed, showing oil tank and the machine gun barrels, exhaust collector ring also visible, A6M3.

Photo A. Lochte.

DETAILS

Above right.
Exhaust stacks and cowling of A6M5.
Photo A. Lochte.

Right.
Two photos of Nakajima NK1F Sakae 13 engine.

Below.
Two photos of Nakajima NK1F Sakae 21 engine.

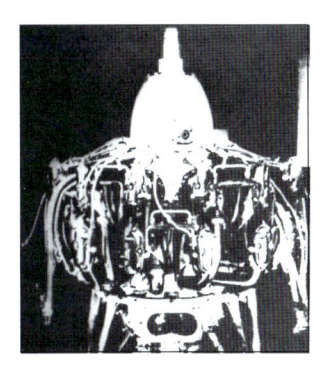

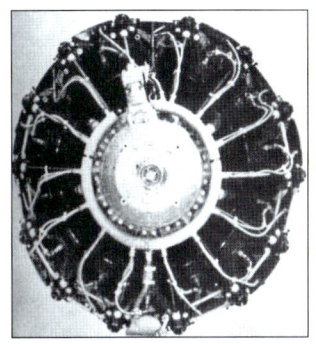

DETAILS

Engine cowling of A6M5.
A. Lochte coll.

Above.
Two photos of A6M5 spinner.
Photos A. Lochte.

Right.
Cowling of A6M5, cowl flaps open.
Photo A. Lochte.

70   Mitsubishi A6M ZERO

COLOUR PROFILES

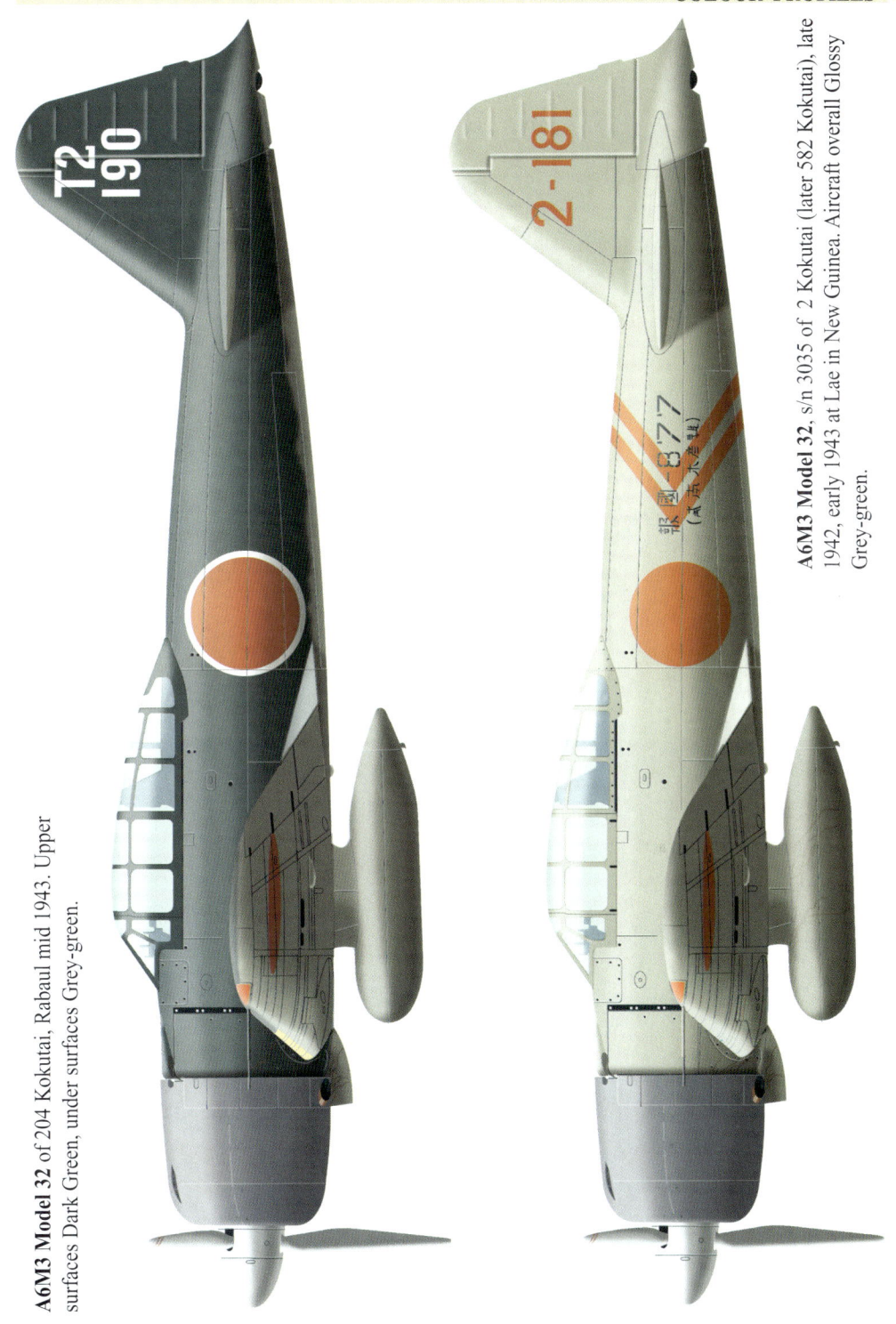

A6M3 Model 32 of 204 Kokutai, Rabaul mid 1943. Upper surfaces Dark Green, under surfaces Grey-green.

A6M3 Model 32, s/n 3035 of 2 Kokutai (later 582 Kokutai), late 1942, early 1943 at Lae in New Guinea. Aircraft overall Glossy Grey-green.

Mitsubishi A6M ZERO **71**

DETAILS

Above.
Exhaust stack and oil cooler inlet, A6M3.
Photo A. Lochte.

Right.
Spinner and propeller of A6M7.
Photo A. Lochte.

Below.
Engine cowling of A6M2 Model 21, front view.
Photo R. Wallsgrove.

DETAILS

Undercarriage

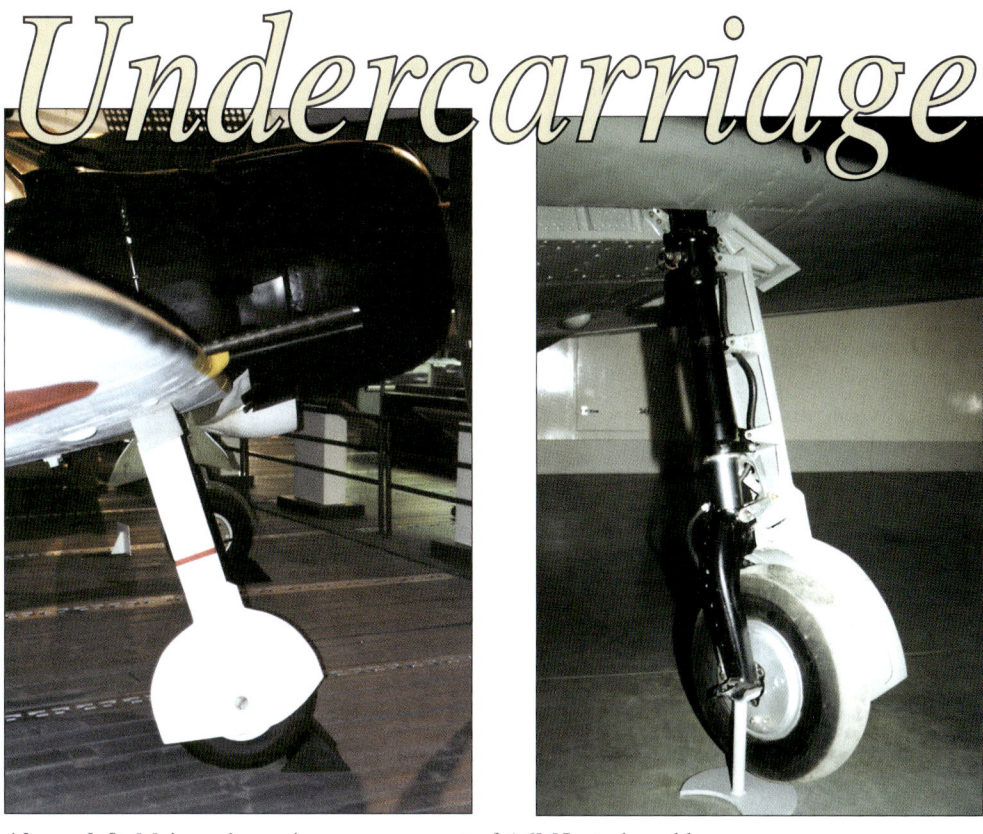

Above, left. Main undercarriage arrangement of A6M5, starboard leg.

Photo A. Lochte

Above, right. Port main undercarriage leg of A6M2 Model 21.

Photo R. Wallsgrove.

Left.
Inside of the starboard wheel well, A6M3.

Photo A. Lochte.

Mitsubishi A6M ZERO 73

DETAILS

74 *Mitsubishi A6M ZERO*

DETAILS

Opposite page.
Top, left.
Front view of the main gear, showing how the doors are mounted. A6M5
Top, right.
Main undercarriage leg cover, A6M7.
Bottom, left.
Starboard main undercarriage leg of A6M7.
Bottom, right.
Port main undercarriage leg, inner view, A6M5.

Left.
Straight view of the port main leg, A6M7.

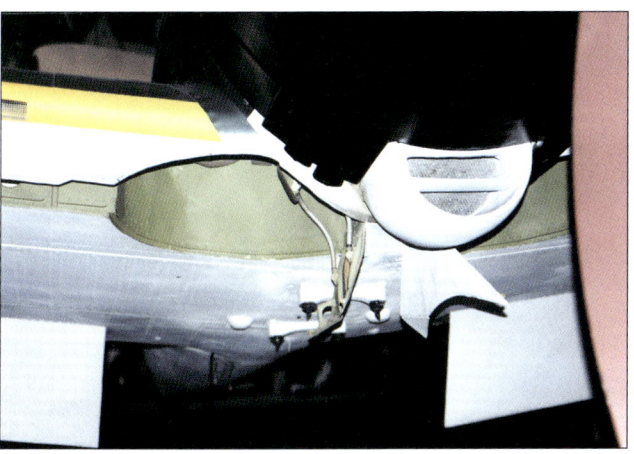

Left.
Main wheel wells and oil cooler. A6M7.

All photos A. Lochte

DETAILS

Right and middle of the page.
Two photos showing upper part of the landing gear cover.
Photos A. Lochte.

Right.
Shot the inside of a starboard wheel, showing tie-down ring and bumper for closing wheel cover.
Photo A. Lochte.

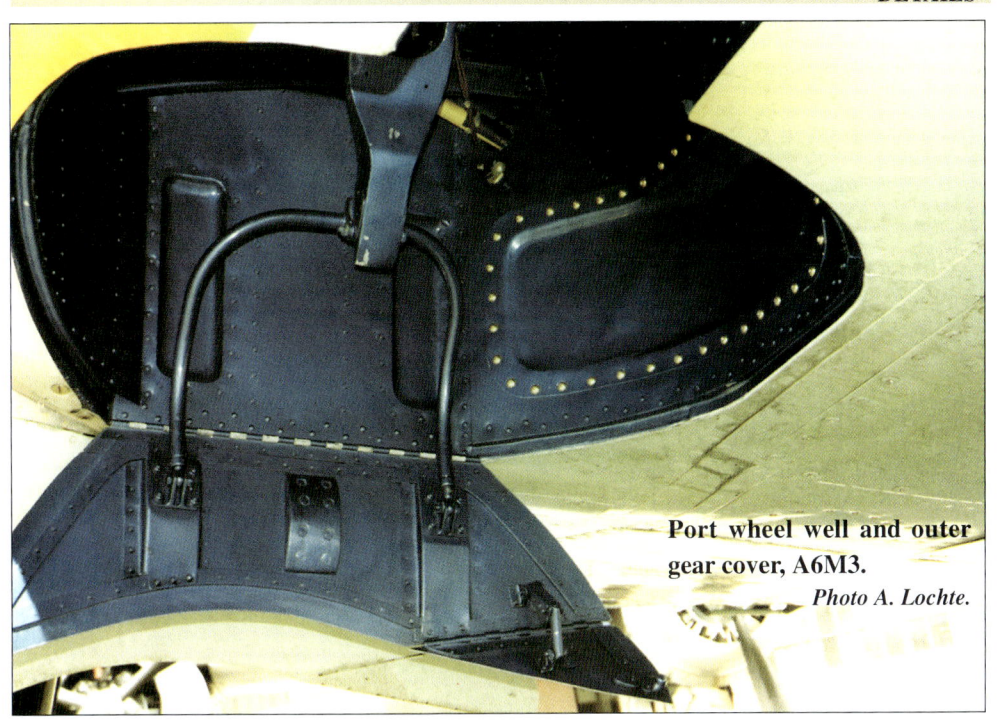

DETAILS

Port wheel well and outer gear cover, A6M3.
Photo A. Lochte.

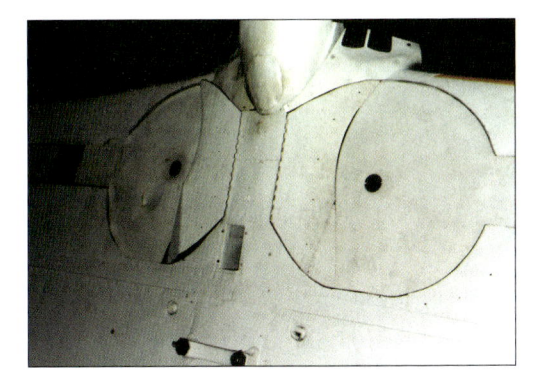

Left. Retracted landing gear of A6M5. The doors' edges are curved.
Photo A. Lochte.

Below, left. Port wheel well, A6M3.
Photo A. Lochte.

Below. Drawing showing arrangement of the main undercarriage leg.

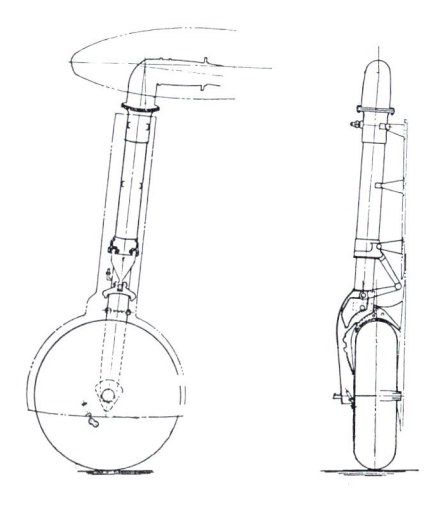

Mitsubishi A6M ZERO 77

DETAILS

Right.
Port wheel well and outer gear cover, A6M7.
Photo A. Lochte.

Right.
Main undercarriage of A6M7.
Photo A. Lochte.

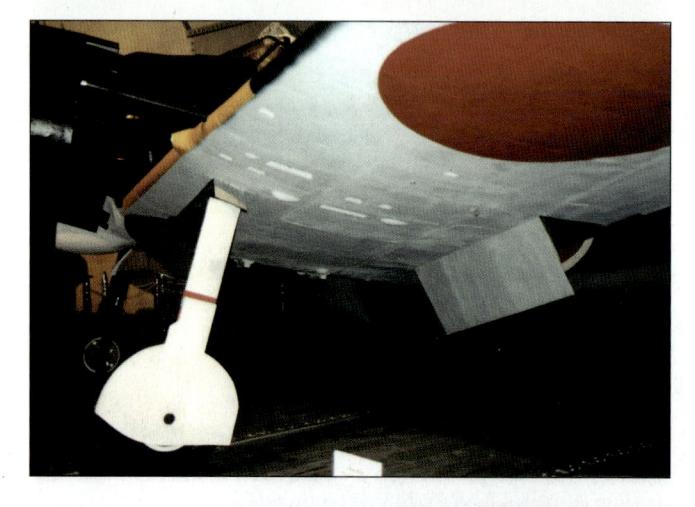

Right.
The way that outer doors are opened is easy visible. A6M5.
Photo A. Lochte.

DETAILS

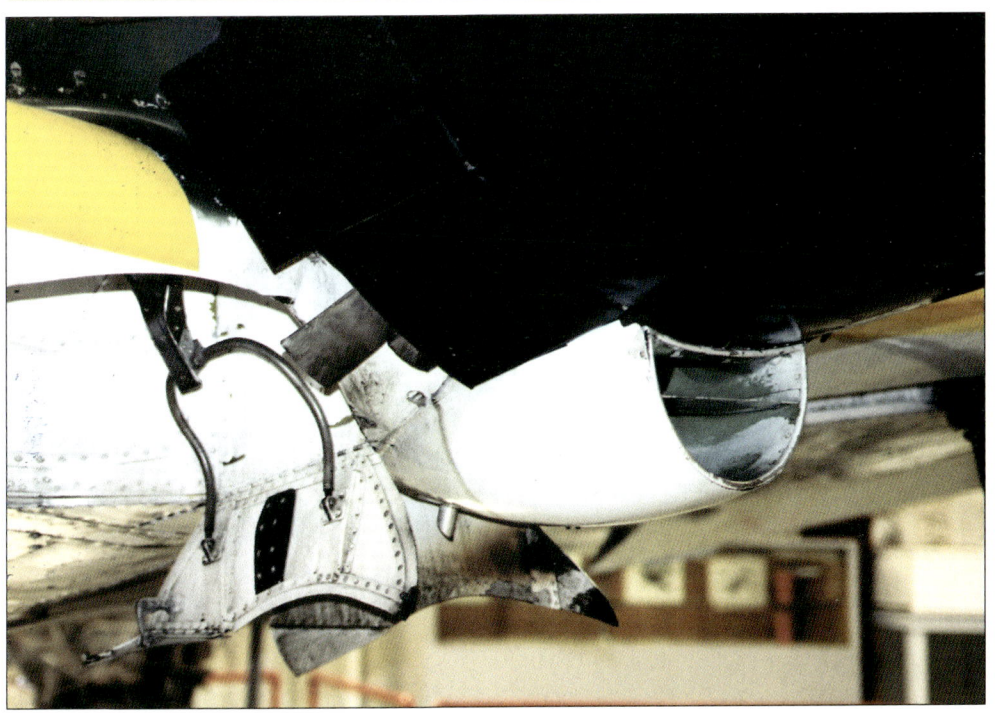

Above.
Main wheel wells and oil cooler, A6M3.
Photo A. Lochte.

Left.
Arrester hook, A6M7.
Photo A. Lochte.

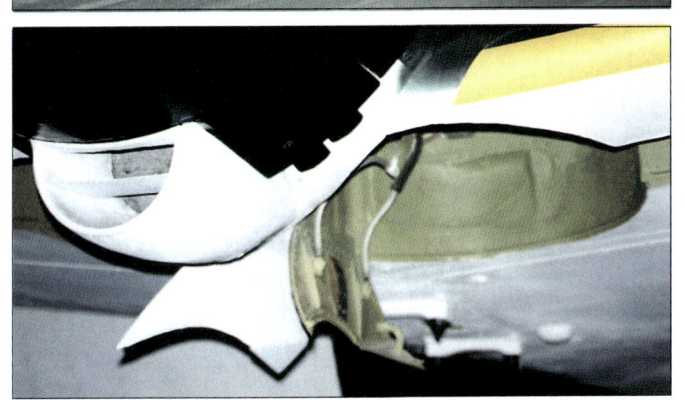

Left.
Main wheel wells and oil cooler. A6M7.
Photo A. Lochte.

Mitsubishi A6M ZERO 79

DETAILS

Right.
Tail wheel of A6M5.
Photo A. Lochte.

Right and bottom of the page.
Two photos of A6M7 tail wheel.
Photos A. Lochte.

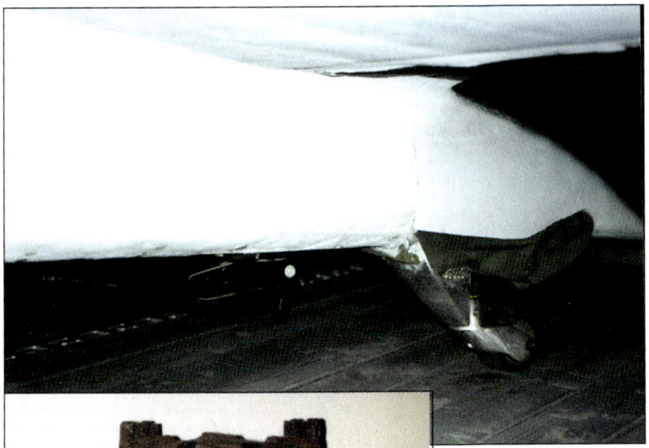

DETAILS

Left.
Arrestor hook and tail wheel of A6M7.
Photo A. Lochte.

Left.
Front view of the A6M2 tail wheel.
Photo R. Pęczkowski

Below, left.
Side view of the A6M2 tail wheel.
Photo R. Pęczkowski.

Bottom of the page.
Tail wheel of the A6M5.
Photo A. Lochte.

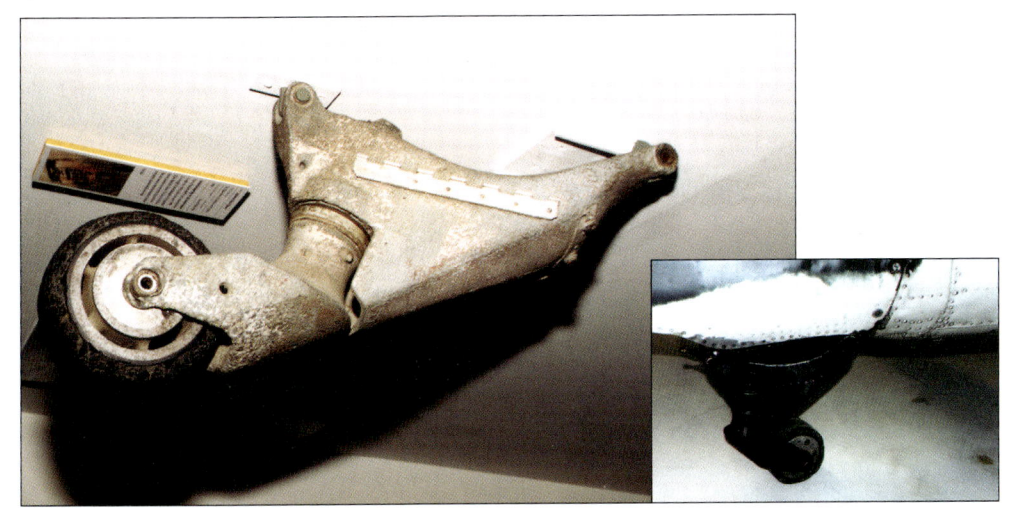

Mitsubishi A6M ZERO   **81**

DETAILS

Right.
Starboard wheel wells, A6M5. Please note all details inside the well.
Photo A. Lochte.

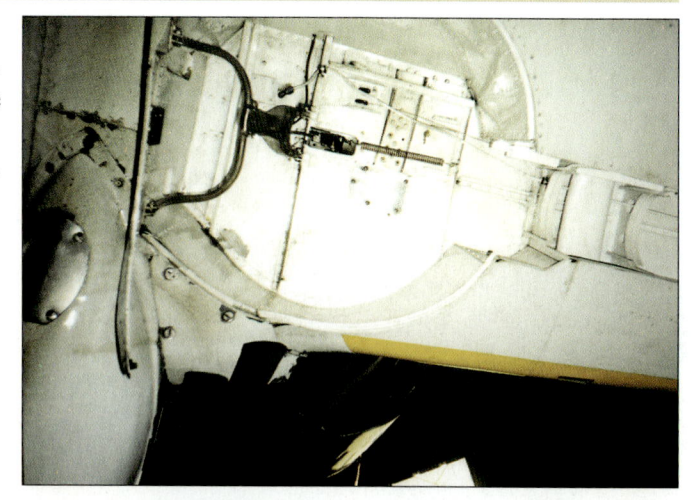

Right.
Shot of inside of starboard wheel, showing tie-down ring and bumper that closes wheel outer door. A6M5
Photo A. Lochte.

Right.
Close-up of the lower part of the main leg cover, A6M5.
Photo A. Lochte.

DETAILS

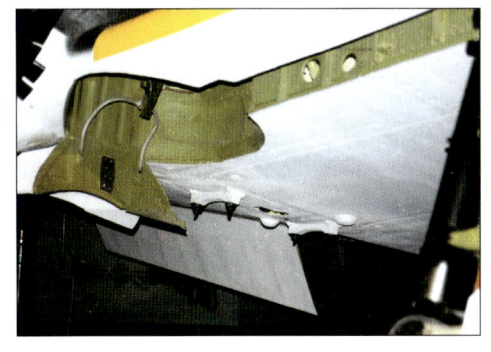

Above.
Inside of the starboard wheel well, A6M3.
Photo A. Lochte.

Left.
Port wheel well and inner view of outer wheel doors, A6M5.
Photo A. Lochte.

Below.
Port wheel well and outer gear cover, A6M3.
Photo A. Lochte.

Mitsubishi A6M ZERO

COLOUR PROFILES

A6M3 Model 32 of 2 Kokutai (later 582 Kokutai), at Lae November- December 1942. Aircraft overall Glossy Grey-green.

COLOUR PROFILES

A6M3 Model 22a of 261 Kokutai, at Kogashima in Japan, 1943. Upper surfaces Dark Green, under surfaces Grey-green

A6M3 Model 22a of 25 Koku Sentai, 251 Kokutai, Rabaul, April 1943, flown by Hiroyoshi Nishizawa. Upper surfaces Dark Green, under surfaces Grey-green.

Mitsubishi A6M ZERO **85**

DETAILS

Armament

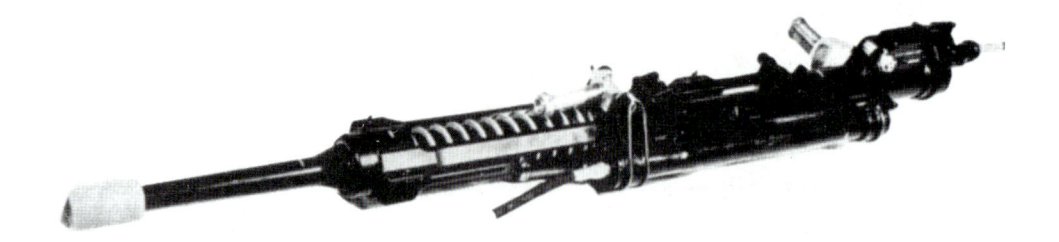

Above. 20 mm cannon Type 99.

Below.
Cannon and machine gun barrels, port wing, A6M7.

Photo A. Lochte.

DETAILS

Left.
Cannon barrel on A6M3.

Middle of the page.
Close up view of the cannon and machine gun barrels, port wing, A6M7.

Bottom.
Once again cannon and machine gun barrels on A6M5.

*All photos
A. Lochte.*

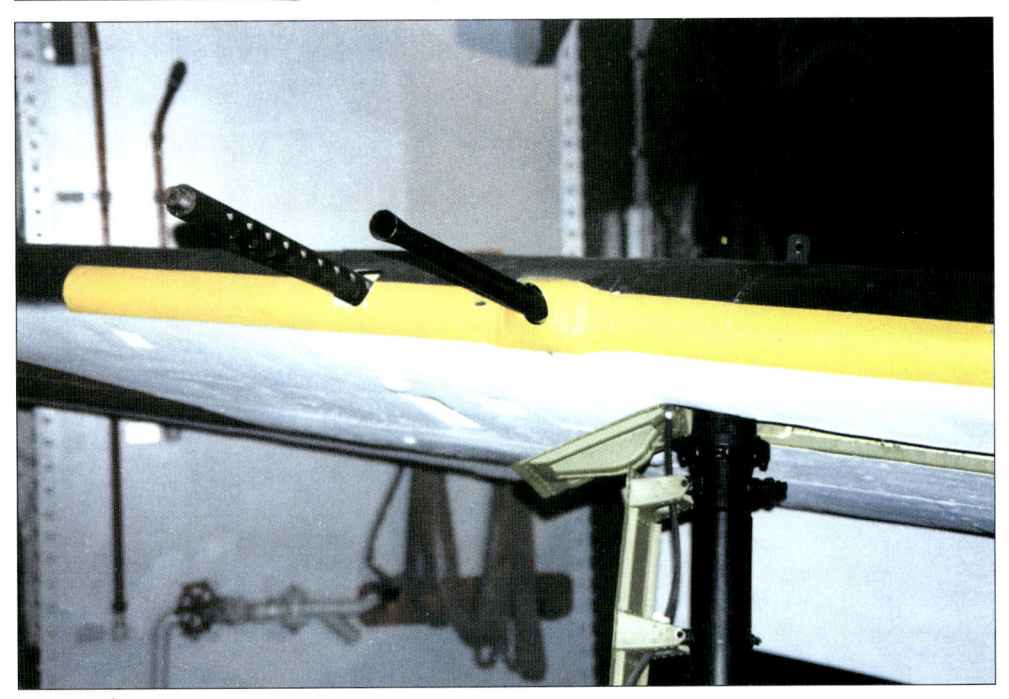

Mitsubishi A6M ZERO

DETAILS

Right.
Machine-gun barrels of A6M7.
Photo A. Lochte.

Right, middle.
A6M3 gun sight,
Photo A. Lochte.

Below.
Machine guns in the cockpit of restored A6M2.
Photo R. Pęczkowski

88 Mitsubishi A6M ZERO

DETAILS

Tail

Left.
Tip of the rudder and fin, A6M3.
Photo A. Lochte.

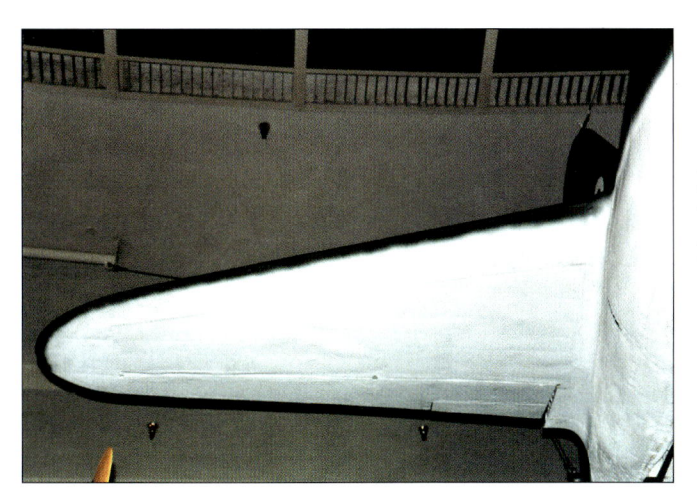

Left.
Underside of the starboard horizontal stabiliser, A6M5.
Photo A. Lochte.

Left.
Port horizontal stabiliser, A6M5. Please note stabiliser-fuselage intersection.
Photo A. Lochte.

Mitsubishi A6M ZERO **89**

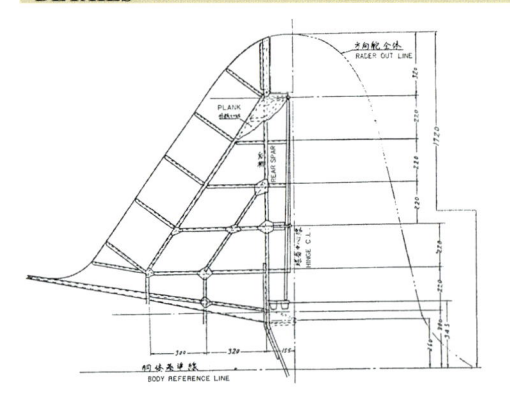

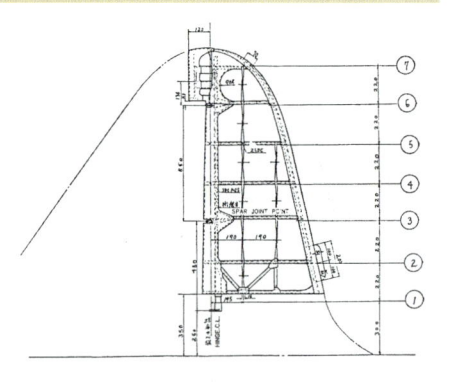

DETAILS

Above.
Dimension drawing of fin. Drawing from Technical Manual.

Right.
Vertical stabiliser, port side, A6M7.
Photo A. Lochte.

Right, below.
Port elevator and trim tab, A6M3.
Photo A. Lochte.

Below.
Tail light and rudder trim tab, A6M3.
Photo A. Lochte.

90 *Mitsubishi A6M ZERO*

COLOUR PROFILES

A6M3 model 22a of 582 Kokutai at kahili Arfield, 1943. Upper surfaces Dark Green, under surfaces Grey-green.

A6M5 Model 52 of 332 Kokutai at Kure, Japan in September 1944. Upper surfaces Dark Green, under surfaces Grey-green.

Mitsubishi A6M ZERO

COLOUR PROFILES

A6M5c Model 52c, Tsukuba Kokutai, Oita, July 1945. Upper surfaces Dark Green, under surfaces Grey-green.

A6M7 Model 63 of 210 Kokutai, 1945. Upper surfaces Dark Green, under surfaces Grey-green.

COLOUR PROFILES

Front and plan view of the typical **A6M5 Model 52** camouflage.

COLOUR PROFILES

A6M7 Model 63. Aircraft with "capitulation" Japanese insignias.

A6M5c Model 52c, aircraft tested in TAIC, RAF. Over-all natural metal with black anti glare panel in front of cockpit.

94   Mitsubishi A6M ZERO

COLOUR PROFILES

A6M5c Model 52c, aircraft tested by 457 Squadron of RAAF. Aircraft was given by TAIU-SWPA at Clark Field. Overall natural metal with black anti glare panel in front of cockpit.

A6M5c Model 52c, aircraft tested in TAIC. Overall natural metal with black anti glare panel only over the cowling and with USAF national insignia.

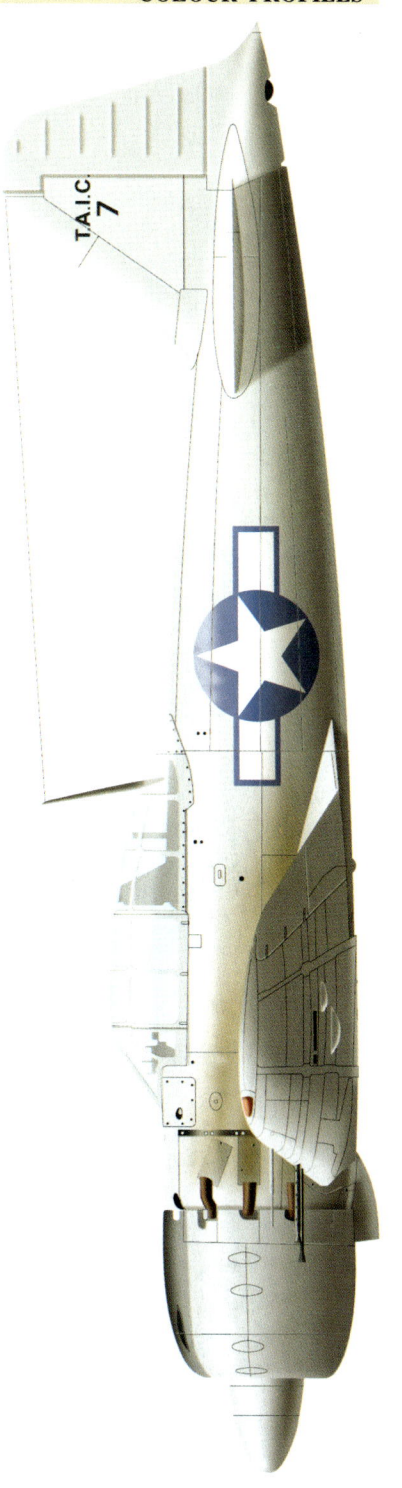

Mitsubishi A6M ZERO **95**

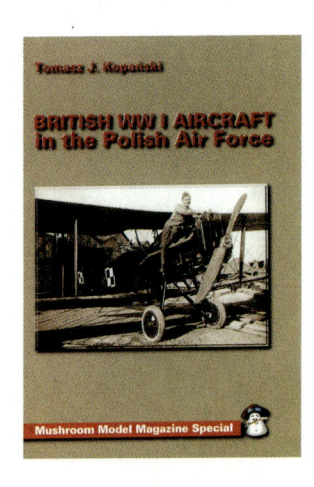

RED SERIES - historical books.
No.1 in the series. The fascinating tale of the British types used by the PAF from 1918-1930, never before told in English.
B5 size with hard laminated cover, more than 100 B&W, mostly unpublished photos. 88 pages include 8 pages of colour profiles - British aircraft in Polish markings.
STILL AVAILABLE

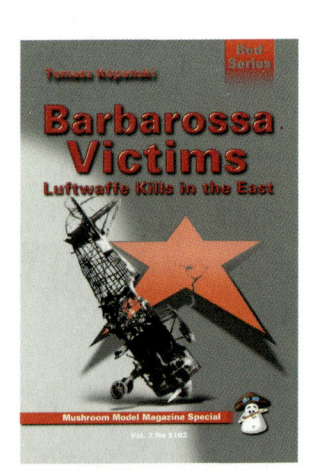

RED SERIES - historical books.
No. 2 in the series. Details the destruction of the Red Air Force during the opening weeks of Germany's invasion of the USSR. More than 350 of B&W, never published photos, Order of Battles tables, maps, glossary. 19 full color profiles, 144 pages, B5 size (SC).
STILL AVAILABLE

YELLOW SERIES
- historical books.
No. 2 in the series. A comprehensive guide to the Bf 109E from the pre-production E-0 to photo-recon E-9. EXTENSIVELY illustrated wITH color and B&W photos (both overall and in detail), profile & detail drawings, and 32! color profiles. Nose-to-tail coverage! 46 color pages, 80 pages. B5 sice SC.
STILL AVAILABLE

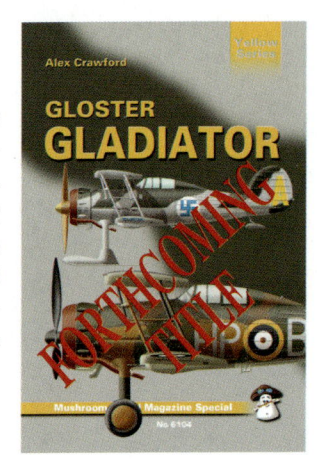